Episodic Toxic Memories, a Survival Story

JOSE MORALES DORTA

ISBN 978-1-950818-63-1 (paperback)

Rushmore Press LLC
1 888 733 9607
www.rushmorepress.com

Printed in the United States of America

THE BRAIN EPISODIC
TOXIC MEMORIES, A
SURVIVAL STORY O22-A

The year is 399 BCE in Athens, Greece. The greatest philosopher the western world has known, Socrates, was forced to commit suicide by drinking hemlock, a poisonous herb. Plato, his brilliant disciple dedicated himself to writing his mentor wisdom through his teachings. Some writers have argued that western philosophy was triggered by the execution in 399 of an old man accused of perverting Athens youth by unorthodox teachings. However, there are legitimate arguments that Greek intellectual world was ripe or developed enough (Tales of Miletus, Heraclitus, Anaximander, Pythagoras, and even the Sophist) to give rise to Greek philosophy and way of life far beyond its geographical boundaries; but none of it had neither the depth nor the influence Plato's Academy has had in the western world. In all fairness, we must add that, either by coincidence, luck or accident, the conquest of Greece by the Roman Empire, kept the Hellenic philosophy, science and art was basically intact until the fall of Constantinople in 1453.

Perhaps it was Socrates' unyielding determination to take up his dialectic method on investigation and argument until the absolute truth is revealed that placed him over two thousand years ahead of his own time. He seems to be an interlocutor of the third millennium arguing the undisputable potential use to humanity of nanotechnology, subatomic particles, black holes, multiple universes as well as the sequence of the human genome for medical purpose. Observation by our senses was not enough for Socrates because it came from the body and the body was unable to attain truth by itself. Life without absolute truth would not have meaning for Socrates. Truth led to self-discovery as it is reflected in your fellowmen. "To hurt or to injustice to another person is worse than suffering it since one is harming one's own soul. Every man must turn all his own efforts and those of his country to bring it about that justice and self-control shall effect a happy life."[1]

Socrates' method of education is not limited by space or time; it has survived over two millenniums the world over and seemingly practiced in academia as well as the general public quite effectively. Following Socrates' teachings, the soul attains to truth only through reasoning...where the soul is alone and by itself, when it strives after truth having neither more communication with the body nor contact with it than is absolutely necessary.[2] Thus, it appears that according to Socrates, absolute truth is attained only by a stainless, not contaminated soul. The body is an impediment for truth to be realized. Well, where does the soul resides in the human body not to be contaminated by it? If man is to attain truth there must be some interaction between his souls and man's body, but the soul seems to loath matter, meaning man's body. According to Socrates, the soul existed before the body and will continue to exist after man's death. What is the purpose of a soul existing in a body that at a minimum instance of contact-contamination- will bring down its pristine celestial purity? The idea of an immortal soul was introduced to Christianity at the beginning of our Current Era. The Roman Catholic Church fathers, among them St. Agustin and later St. Thomas Aquinas, re-affirmed on Christians that the soul had a divine origin. Under such church dogma, I suspect nobody in its right mind would dare to question it without risking being burned down on a public square. We in America were not free from these witch hauntings. Religion intolerance was not restricted to Roman Catholic only. As matter of fact, religion intolerance has caused more death and destruction than all the wars I know of. René Descartes of the Jewish religion, possibly following Socrates, claimed that human beings have a double nature; one is made flesh, blood and bones, and a mind that comes from the divine soul. These dualistic natures of men still persist in many

---

[1] The Dialogues of Plato in Gorgias, part 111, p. 358, Banton classic, 5/2006, random house, N.Y., N.Y.

[2] Ibid, Phaedo, p. 81.

quarters of western society. Charles Darwin with his theory of the Origin of Species, the Double Helix, Dolly the Sheep, Recombinant D.N.A. and Craig Venter creation of a synthetic bacterium have not simplified things for us.

It seems to us that a physical and mortal body is a prison and prisoner with no ultimate goal, but to exist without ever attaining its independence. However, this homo-sapiens developed a sophisticated brain and has become self-conscious, and is demanding its freedom. Interesting enough, about two thousand years after Socrates forced death, Rene Descartes in a similar search to attain truth, chose the pineal gland deep in the brain as the locus where the soul and the body interact. At present time we are not in a race to unravel the mysteries of the soul, if it truly exists, but our aim, among others, is to find out how the brain communicates with itself and creates our wonderful mind in each of us. However, despite our most recent scientific advance during the twenty century and first decade of the third millennium, Greek religious philosophy stands strong in our society. While Socrates used the method of answering a question with another question; he seems to have used a relatively modern scientific method. The search for the absolute truth is the ultimate goal and you must put aside prejudice, input from our senses or personal and family ties and preferences. Cause and effect, a basic dogma of science, seems to be paramount for him as well as present day scientific research.

Rene Descartes made use step by step methodology to achieve truth by attempting to rule out or eliminate any doubt while pursuing truth; he would eliminate the doubt that was tainting or blocking truth by addressing the doubt. It seems that if he eliminates doubt by self-reasoning, when doubt no longer exists in his mind, when there was not the slightest notion of doubt in his mind, truth would be achieved. It seems that Descartes was engaged in an unending self-scrutiny, self- questioning to achieve self-understanding. The thinking self (cogito ergo sum) was his brain child. If there is a cloud in all human knowledge, Descartes thought he could disperse it or even eliminate that cloud through self-scrutiny and empirical data provided by rational thinking residing deep in us. To some of us,

Socrates Phaedo was deeply inside Descartes. If for the sake of argument, we take the "soul" of both, Socrates and Descartes, out of our sight and focus our intellect on "cogito ergo sum" as the savior of mankind, we will see our humanity. Cogito by his thinking power could win independence and freedom from a soul that its sole purpose is to go back to where it came from. The soul despises the body as a contaminating blob of dirt that a minute contamination would taint her and delay her journey back to an angel or a god status. Descartes Cogito's power resides in his ability to independently think and gain self-consciousness. This child or individual entity of Descartes, cogito, through self- awareness could explore, study, analyze, plan and master his environment; the environment that he needs to able to survive independently from the soul. Cogito could be more powerful than many souls, especially, underdeveloped souls that are unaware of their own identity and source of origin. Cogito could be the liberator of the human race. He could reduce, enter in to a contract or even eliminate all ties with the super-natural whatever it might be. However, for cogito to become his own master, his observations have to take the next step in evolution, the scrutiny of his observations have to be subjected to verification through an impartial scientific method. Cause and effect must reign over Cogito's world to sustain itself and humanity in general.

We can argue that Descartes' rational and self-cleansing man is the ultimate goal of any soul. The soul is an extra-terrestrial entity that inhabits the body of earthly animals, including men. And, of all the animals on planet earth, the soul seems to have chosen homo-sapiens as the most viable candidate to develop self-consciousness and master the world for his benefit. Thus, we may conclude that man develops into a soul ambassador on earth. The soul may want to assist man achieve its independence and freedom from his own mortal-physical body and become master of his world. The soul has to keep its distance from man or material world to maintain her purity and freedom to travel at will. We may ask ourselves what is the soul's purpose in assisting a corporeal man achieve freedom from himself. A possible answer to this question is that the realization of Cogito as the liberator of a flesh bodily tied man would justify her existence.

Once more, we are going to make use of a crutch to be able to maintain our argument, and maintain your focus on our essay; there are two atoms of hydrogen and one of oxygen joined together forming a molecule of water. Every plant and animal on earth needs water to survive. Water, along with other organic molecules gave rise to all form of life on earth. We can see the beauty of a water cascade, the beauty of a sundown in the horizon as well as the beauty of flowers in a garden. However, despite all this beauty, the atoms that made beauty possible lost their independence and freedom while trapped in a tree hard bark or animal body full of billions of bacteria. However, had not they had joined or bonded to each other, neither hydrogen nor oxygen could enjoy the beauty of their own creation. Plant captures energy from the sun and converts it into starch and, we animals feed on plants for our survival. During this simple analogy, the atoms of hydrogen and oxygen can claim that the wisdom they gained during their pilgrimage as men and flowers over compensates for their loss of individuality as individual and separated atoms. The soul may make the same claim. The soul extends her-self as a guiding principle while the rational-thinking man masters his own body, and his own world. And, in addition, he will be capable of contemplating and admiring his own creation. Man becomes immortal through self-scrutiny and reflection over his creation. The thinking man-cogito- creates his own mind capable of traveling in space and time while still residing on this planet, earth.

FEARLESS SOCRATES BEFORE HIS INCOMING DEATH

Socrates was not afraid of death. His soul had taught himself discipline without being tainted by Socrates' bodily needs and desires. Socrates became immortal. He became immortal by his teachings through Plato, but Socrates' body, especially, his brain, an organ that generates electricity never disappears. Socrates' electrical energy, never created nor destroyed, is also immortal. Who among you can claim that Socrates' psychic energy has not touched you and assisted you understand yourself? There are countless cases in all walks of life (social, economic, educational and religion orientation) that claim to have had an experience and work done under the assistance and presence of some source of energy unable to describe. Homo- sapiens have a very sophisticated brain with at least one hundred billions neurons with the potential of making trillions of synapses. A synapse is a space between two neurons. It is the place where communication between brain cells, known as neurons, takes place. It is also, according to most neuroscientists, the location where most learning occurs. Actually, communication at the synaptic level is a chemical triggered by an electrical impulse that originated in the axon of the neuron by sodium and potassium pumps.

Socrates, Plato, Aristotle, Da Vinci, Picasso, Galileo, Columbus, Descartes, Napoleon, and even Adolph Hitler's brain created electrical energy. We know that the basic unit of any element is the atom. All animals and plants are made of atoms, molecules, and cells. Four elements (hydrogen, carbon, nitrogen, and oxygen) make up about 96.5 of our bodies. According to both, Socrates and Descartes, the soul is incorporeal, meaning, that it has no body, it is not made of matter. Atoms, the building blocks of all molecules and cells have matter. In the nucleus of every atom, you will find protons, neutrons, and electrons. But, even that tiny particle, the electron, has weight and is matter. The soul, therefore, cannot be found in protons, neutrons, and electrons of an atom. The soul, not having matter has to avoid being trapped by it if she wants to go to her celestial paradise if that is her ultimate goal and absolute truth. The soul's trip to earth to inhabit a man's body is a complicated endeavor. If she is an absolute truth, an impeccable, spotless, clean of any impurity, the mere presence of an earthly body is an abomination. She might see it as a punishment since any contact with matter is to be unclean, impure and degrading. Momentarily, the soul may go into self-denial to be able to interact with matter in a man's body. If she is tempted to this, she may lose her right or privilege as absolute truth. The truth carrier, the soul itself compromises her own existence as a soul capable of breaking her ties to a mortal, be it Socrates, Descartes and you.

So, what is the nature of a soul? Secondly, where would the soul interact with men without being contaminated herself? The soul might be seduced by man's created beauty on earth like gardens, boat riding, music or even his power to subjugate and enslave other animals. However, the moment she feels the slightest enjoyment in this earthly and passing pleasure, she will become clouded and unable to see herself as she has become part of man and the world he put together from matter. She might even fail her trip to earth if she does not find someone like Socrates who was willing, against all odds, to go out in the pursuit of truth. I remember my father saying over and over again, "Search deep in you, scrutinize every thought and feeling that touches you and you find the truth that will set you free." I never asked him, but I was wondering, set me free from what? Once more, we repeat that the soul's place of residence in man's body is highly questionable; we concluded that it cannot be in cells, molecules, and atoms because they are composed of matter.

THE SOUL AS A CATALYST AGENT IN A HUMAN BRAIN

We shall go to the next stage or phase of our search for the location of our soul in our brain. We will explore the soul as a catalyst agent rather than a direct doer of action within a man's body. A catalyst agent is capable of bringing about changes without suffering any change itself. Eureka, we have come to a fertile ground; there are many catalytic agents in our body, namely enzymes. However, enzymes are proteins and proteins are made of amino acids. Proteins are the building blocks of your body. From head to toe, we are a big mass of proteins, in other words, matter. However, there is a possibility that we might find a location in our body and it is not going to be the pineal nucleus of Descartes. We have billions of cells that, although made of matter, produce a non-matter component, namely, photons. Axons, the long processors in neurons are the site of electrogenesis of our nervous system. There are around 100 billion possible sites for creating electricity in our brains.

Each brain cell has a body, or soma, with many dendrites that receive information from other neurons. Also, extending from the body is a long process or fiber called axon. The axon is connected to the cell body by a ring named the hillock. The hillock plays a major role in deciding which impulses coming from the cell body pass into the axon itself. In the axon, we will find sodium and potassium pumps which, among other things, are responsible for creating the non-organic or material part of our brain and body. We are assuming that the soul would not have objections communicating with our perishable- corporeal body through the electricity produced in our brain cells. As far as we know, electricity is a non-contaminating catalytic agent capable of fast and long traveling. It can travel through matter as it does through copper without losing a significant portion of its strength or producing any contamination. Axon electrogenesis is crucial for our survivor. Most of us are aware of one of the most fearful and degenerative maladies afflicting mankind, Alzheimer's disease. The Alzheimer patient loses his memories and consequently, his self- identity.

As the disease progresses and neurons lose their ability to produce electricity, thus unable to send a chemical message – a neurotransmitter - into synapses there is no communication between nerve cells. If there is no communication between neurons because axons no longer can produce electricity, different regions of the brain collapse. If there is no communication among brain cells, we cannot retrieve or store our memories and ultimately, we lose our self-consciousness and self-identity. We not only lose consciousness and self-identity, but also the ability to move, walk, talk and do for ourselves. In Alzheimer's disease, there are plaques between neurons and the internal microtubules of the neurons become entangled making it impossible to perform and function normally. As you look inside a neuron, you will see multiple organelles traveling or moving along the cell body and axon doing specific and complex jobs. These are very specialized proteins that become dysfunctional and collapse the entire nervous system. At the present time, the disease is idiopathic, meaning that the cause is not found yet, although in about 10% of patients there is a group of genes involved. Besides the organic component of the disease, there are environmental factors as well as causal factors developed during pregnancy.

MY SCHOOL TEACHER, DELGO

But let's go back to the soul and its contempt to mix and communicate with our body because it is corruptible, ephemeral and an obstruction to gain absolute truth. During my school days, I had great difficulties, meaning problems, understanding how a soul could interfere with my learning and behavior in the classroom as well as getting along with our classmates. My science teacher, Mr. Delgo used to teach us that the smallest particle an element could be broken into without losing its identity was an atom. But, an atom is composed of protons, neutrons, electrons, and other smaller subatomic particles. I used to argue that the soul, as far as I understood it, was a very picky and discriminating lady. Perhaps I was identifying the soul with my mother. The soul would not pick an atom as the locus for contact with the body because the atom's protons, neutrons, and electrons, no matter how diminutive they are, are still considered matter. It was not until later on in college that I learned that photons and other subatomic particles do not have matter. However, at the time I was obsessed finding a place in our body, especially, in our brain that could be acceptable to the soul. I strongly believed that I had a soul. I used to refer to Socrates for all references regarding my curiosity as a soul seeker. My teacher used to punish me for insisting on finding a place for my soul in my body. I recall that he used to call me to stand up in front of the classroom and point to my shoes, which most of the time were dirty on account of the dirt road I had to walk to get to school every day. And, sarcastically, Mr. Delgo pointed to my shoes and loudly says: "there on your shoes is your soul."

Mr. Delgo's abusive behavior, that is how I considered his unjustified action towards me, was reinforced by similar experiences by some of my neighbors. However, following Delgo's abusive behavior, some boys and girls used to claim that my soul was a wicked one. They used to point to a cave with an underground river as the source of my soul. A couple of youngsters around age twenty had left our

EPISODIC TOXIC MEMORIES, A SURVIVAL STORY

neighborhood and most neighbors pointed to the cave as the culprit for their disappearing. At times, when I was upset with my teacher, my mother or some neighbors, I used to go to the cave with a home-made torch made of cupey resin. An old, but loving Taino Indian lady had taught me how to make torches with cupey resin. She used to claim that she learned it from her parents who did not have money to buy kerosene to light their home at night. I used to love her telling me Taino Indian stories, and to encourage her to entertain me with her beautiful stories, I brought her fruits and vegetables from my father's farm. The time I spent listening to her stories have become permanent memories in my brain. I often go back in memory time travel and enjoy her company and love as if it were in the present time. These are culturally rich and lovely memories inside my brain that have become part of me. In retrospect, I believe mama Tiny, the Taino old lady, as many of us called her, was providing for my mother's missing love and closeness. My mother had eleven children and her child immediately before died when he was about 2 ½ years old. At the time, around 80 years ago, I believe nobody knew the cause of his death, but I suspect, from her description of his symptoms, that he died from the thalassemia disease. It is a disease found among Italian and Greek people. I have thalassemia trait as well as three of my siblings. My mother spent most of her spare time, if she had any left, caring for ten children besides helping my father look after the ladies that used to come to work during the tobacco harvesting season or other minor crops.

My teacher's hard discipline and punishment towards me for my insistence in allocating a place in my head for my capricious soul have always haunted me. I became a very shy and anxious person suspicious of teachers and people in authority. When I had to stand in front of an audience my knees used to shake and often click against one another despite my efforts to control my anxiety. The image of my teacher, Mr. Delgo was always present each time I had to address a group of people. I avoided a group of people thinking that I may have to address them in one way or another. Meanwhile, I could not let go of the soul. I used to avoid listening to preachers because of the emphasis placed on the salvation of every human being soul.

Mr. Delgo used to punish me for pursuing a locus for my soul and body-safe communication while my neighbors pointed to a cave for the origin of my soul while Socrates gave it a place in heavens. If this thought-provoking riddle is not confusing for a 15 years old country boy, nothing else is. As a growing youngster, I felt alienated, ostracized and justly rejected by neighbors and friends because I was looking for something most people in the neighborhood did not care to question or knew anything about it.

MAMA TINY AND BIRDS FLYING IN THE SKY

In retrospect, I wonder if I was not, in part, responsible for, or contributed in any way to my isolation and suspiciousness of my neighbors about my questionable behavior, both at school and at home. It seems that my neighborhood was too small for me. Besides my unending quest for a conciliatory meeting place for my soul and body, I was intrigued and almost obsessed by birds flying in the open sky while I was tied to the ground. I got stuck with small jumps that could hardly get me anywhere. I used to climb trees not only to eat and enjoy their delicious fruits, but to show off my ability to climb up on top of it and slide down the branches skillfully and impressively. I recall how our neighborhood's boys and girls, as well as few adults, used to gather around while I was sliding down on a tree. I was a small and thin boy relative to my age, and truly enjoyed people watching me showing my acrobatic skills. Mama Tiny, who kept an eye on me, was blaming my neighbor's kids for me showing off up and down the trees. When she came around and ordered me to stop and come down the tree I always obeyed her. If I sensed that she was upset or angry towards me, I would go out and bring her fruits. She usually warned me that I could not buy my way out by bringing her presents. She used to say to me, "stop pretending you can jump and fly like a bird on the trees, you are going to get hurt yourself." Well, that did not stop me from doing it again; on the opposite, her concern for, and her defense of me despite young and adult demand for more of my show off acrobatic skills, only reinforced my behavior. I had an applauding audience and a lady, mama Tiny, whose love for me made me feel great and happy. To make it a little more interesting but extremely painful for me, my idea that I could fly like most birds, seems to have had no limits. My mother used to raise many chickens and turkeys at home. In fact, there was a big ranch close to our home for chickens and turkeys only. My mother's income came from raising and selling chickens, turkeys, and hogs. My father used to take care of the income coming from the cattle, horses, tobacco and major crops pocketing all the money. In our ranch, there were 100 empty pounds of wheat, corn and fertilizer bags. I used to watch how turkeys attempted to fly but hardly could get off the ground. They were too heavy for their small wings to carry them up in the air. It occurred to me one day that I could make wings out of empty bags from the ranch. Attempting to convert my genial idea into reality, I went to a nearby bush and cut two thin and light twigs and attached to it an empty bag cut in two pieces. Each wing or half-empty bag was held by my left and right hand, extended outwards and attached by a string taken from the same tree to each of my toes. My idea looked great. I could move my hands and legs freely. I could not try it on mango trees because I needed an empty space to jump from.

JUMPING FROM COCONUT PALM TREES

My problems and luck have never failed me. There were three small coconut palm trees about half a mile from home. Surrounding the palm trees there was thick and tall grass that in case of a rough fall, it would serve as a cushion for me. I climbed up the palm tree and waited for a strong wind to come and jumped off. Of course, I did not fly, but I glided down the tree and landed a few feet away from the tree trunk. The grass on the ground proved to be an excellent cushion. I had slight bruises, but I felt great, I had accomplished one my dreams, to fly. I felt tempted to do it again immediately after I landed on the ground and I wiped out any dirt or dry leaves attached to me. However, the sticks that I had used to attach the empty bags to my feet and hands were broken. I must have been soaked on dopamine because, despite a little bruise here and there on my body, I did not feel it. I was so happy that I wanted to run home and tell my experience to my brothers and neighbors. However, I was fearful of punishment by my parents and disapproval from neighborhood adults. I only talked about it to a kid that I could intimidate with aggression if he did not keep our secret. He even agreed with me to come to the palm trees and watch me fly. Well, I repeated my debut four times with great success. My friend wanted to try it himself but when the hour came to climb up the tree and jump down, he became very frightened and chickened out. The news that I was flying spread in the neighborhood like fire. The oldest son of our neighbor, a landlord, had seen me jumping from the palm tree and had gone directly to my father. When I saw him going to my father, my pants got wet. My legs began to ache before the whipping had begun. His word was enough for anybody in the neighborhood to take disciplinary and appropriate action. My father was waiting for me to get home and without a word, he grabbed me by the hand and gave me six lashes.

Well, that put an end to my flying wings. My skinny legs were showing the slashes by swelling and some of my less friendly neighborhood boys used it to tease me pointing to my skinny and painful legs. Deep within I was happy; I had proven that I could glide if not fly. That was not a small thing to do for a country boy over 70 years ago.

This experience took place around 1938 and I can appreciate and love this memory very vividly in my brain. I may add that had I not had a soul, perhaps I would have been deprived of this rich and beautiful memory. However, I must tell you that this is not the only thought that haunts me. My soul did not stop in pushing me or at least motivating me to do the unusual, the uncommon and perhaps ridiculous things. Thus, making people laugh and enjoy themselves while accusing me of introducing dangerous games and ideas that may have hurt their children. Many times, I heard adults, but particularly parents say, "We should close that cave, it has an evil spell. That cave must be bewitched; it is the place where he gets those devilish ideas." At the time I did not think of anything else but keep myself busy. My mind or perhaps my soul was always thinking of something unusual to attract the attention of youngsters and adults from the neighborhood. I wanted to leave the ground and fly or walk above the dirt that the soul despised so much. Therefore, I built a couple of stilts and practiced walking on it before I took it to my oldest brother's grocery store.

My brother Emmanuel store, a social magnet for all of us

My oldest brother Emmanuel, was a brilliant young man who was taking corresponding courses from a school in Chicago. He had a small grocery store, the preferred gathering place for the whole community. He had a radio run by batteries. People used to come to him not only for groceries but to learn from him about things that were taking place around the world. The evening that I presented myself walking above the ground on stilts is indescribable; some adults were laughing and applauding me, while others wanted to take my stilts away and break it. I had brought another curse to the neighborhood and the boys would be trying to imitate me and get hurt. I recall that I went around my brother's store window and looked inside. My brother looked at me in disbelief and walked towards me, but before he could take a good look at my stilts, I went in front of the store where a few people were already enjoying my new attraction. My brother came out of the store, briefly inspected my stilts, and with a broad smile on his face, said to me, "make sure you nail the holding section well." It meant to me that the smartest person in the neighborhood had approved my showing off. I was about thirty inches above the ground and made me one of the tallest persons there. Well, the fun did not last too long, within a few days, there were boys attempting to do the same thing and some got hurt. Word got to my father and he demanded that I bring him the stilts so he could apply the machete treatment and finish my new attraction. It seems that no matter what I tried to do to entertain myself and neighbors, I ended up almost like Socrates, perverting the boys in my neighborhood. It is up to you to pass judgment. Would you blame me, my soul, the neighborhood or none of the above mentioned? Mama Tiny was always there for me and after giving me a cup of black coffee, she said, "My child, no matter what they say, you will find something new to amuse all of us. Young and old, we love and enjoy the things you do, just be careful."

My brother kept in touch with news around the world through his radio and newspapers he bought in the city. World War II was coming in soon. Hitler had invaded Austria and part of the Czech Republic. Neighbors used to get advice from my brother regarding whether the United States had to enter the war in Europe. Many parents were afraid their sons would be called into military services. Some gossipers used to say that Adolph Hitler had no soul or that his soul belonged to the devil. My preoccupation with my soul was attentive to any comment regarding anyone's soul. My brother, as well as neighbors, described Hitler as a criminal individual whose aim was to proclaim his race superiority doctrine around the world. We were afraid we would become guinea pigs and slaves. Moreover, the swift victories of Hitler's army made us scared. When my brother announced the fall of Paris to Nazi military forces, adults took their hats off and maintained a moment of silence in prayers. My brother had only gone to ninth grade school but was well-read and kept many books at home. He kept us well informed about cities like Paris, Madrid, London, Moscow, Berlin, and Washington. Also, names like Churchill, Roosevelt, Stalin, Rommel, Patton, Mac-Arthur and later, Timoshenko and Zhukov in Russia were household names at home. At the time I repeated their names as if it were a game, but later on, in my adult life, I studied it in depth. General Patton, Rommel, Zhukov and Bradley became my idols for years.

MY BROTHER JOINED THE U.S. ARMY, A SAD MEMORY

My brother Emmanuel had been sent to fight the Nazis in France in December 1944. The 65th Infantry Regiment came under Patton's Army. He became one of my heroes for life. I considered him the most brilliant American military officer comparable only to Robert E. Lee during the civil war. Rommel was a Prussian career soldier; I enjoyed his unparallel skills in tank warfare. I enjoyed Rommel even more as later on I learned that despite his love for motherland and discipline as a military officer, he was involved in a plot to assassinate Hitler. In December 1945 we learned that Patton had died as a result of a car accident, I wept next to my brother, Emmanuel. We blamed the Nazis for his death. We ventilated our anger towards the Nazis, even more, when we learned the concentration camps and the death of millions of innocent people. I recall my brother calling Hitler the evilest individual in human history. He used to tell me that even Attila and Genghis Khan were minor figures compared to Hitler's thirst for blood. He further commented that Hitler had no soul; he was an incarnated fallen angel like Lucifer or Satan. His observations about the Nazis and Hitler in particular, seem to have motivated me further ponder and search for my soul.

The year 1945 was coming to an end. I enjoyed my brother's talk about Europe and his experience during WW11. He used to describe Paris as the most beautiful and enlightened city in the world. I do not know how or where he was able to learn so much about France's history and culture. His world knowledge was beyond my comprehension. Incidentally, he had a very good command of Greek and Roman history. He had read Plato's Dialogue and occasionally used to touch on Socrates' search for truth. He could read and understand English fairly well, but I could not. He had some books written in English but I was not interested on it at the time. My curiosity was the soul. In retrospect, it seems that the soul that I was looking for was not an abstract entity or thought, but rather something I could hold and restrain. If I was persistent and determined enough to go after it despite its elusive nature. Well, my school years were coming to an end. It was June 1946 and I was very glad to say to Mr. Delgo: "Hasta la vista, baby." The memories of humiliation at his hand had become permanent and recurrent episodes in my life and popped up spontaneously punishing and torturing me for so many years. I was already 16 years old and weighing around 75 pounds. Mr. Delgo was six feet tall, 180 pounds man; there was no match between the two of us. I just wanted to be out of school and away from him.

AWAY FROM HOME AND INTO THE CITY IS ANOTHER EPISODIC MEMORY IN MY LIFE

My neighborhood was becoming too small and boring for me despite my brother's vast wisdom and world knowledge that captivated me. By Christmas 1946, I had left my father's home and moved to the city. I found a job as a clerk working twelve hours a day, six days a week. Soon thereafter, I moved to my third oldest brother's home. He was seriously looking for the soul. He had joined a worldwide group with local chapters in New York City, London, Paris, Madrid, and Buenos Aires. Soon, I learned that he had taken another route to get to the soul. Unlike my mind travel, flying from trees and Mr. Delgo's punishment, my brother had many books in his home that I could read. He was very much interested in parapsychology, metaphysics, Hindu religious philosophy, Christian religion as well as spiritualism. I thought I had found a gold mine that would lead me straight to the soul I was looking for. At his invitation, I attended a few spiritism séance but was not impressed by what I thought was a place to ventilate your daily problems. I used to have very interesting and productive conversations and arguments with my brother over my search for my soul. After all, in essence, he was a Christian believer that liked to quote the Old Testament prophets to support his arguments. For him, man's essence was his spirit that came from G-d. At the time, I did not equate spirit and soul as the same. I guess I have never had a clear picture of one and another. My brother had both, the Roman Catholic and Protestant bibles. I could not understand a G-d that took sides in a war; became authoritarian and angry and punish people for things he must have known they would do. G-D was omnipotent and omnipresent. He knew everything in advance. As I read the Bible, I thought men would always make mistakes and should not be punished by G-d. If a man breaks any rule he should be punished and appropriately taught by the people he lives with. My brother liked to say that whatever wrong you do during life on earth, you would be accountable for during this lifetime or the next. I tried hard to understand my brother's argument but was not satisfied with most of his conclusions. One wonder if my teacher's sour memories interfered or had anything to do with objections to what I thought was a capricious G-d as well as my brother insisting that any wrong done by us had to be addressed in this life or the next. It did not ring well in my ears.

MR. CLEMENS, ANOTHER AUTOBIOGRAPHIC MEMORY

There were a liberal thinking and bright person that had a small shop near my brother's house. Let's call him, Mr. Clemens. Occasionally, I used to bring him my torturing memories and struggle with my soul. I guess he liked to listen to my stories if it is stories. I like to call it my life history instead. When I was walking on the opposite side-walk to his shop, he used to call me in and asked how I was doing with Mr. Delgo. We both enjoyed a big laugh, but in reality, I wished he never brought his name to memory. I wanted to bury Mr. Delgo as deep as I could, perhaps throw him in the cave so the river might sweep him away and into the ocean. In a way, I was happy every time he called me in. He had lived in New York City for over two years. Later on, I learned also that he had gone to Paris to study medicine. He thought I would benefit if I travel or migrated to the United States. He was a well-read man. He suggested that I read a book by Albert Einstein on physics. Of course, that was out of reach for me. Imagine a ninth-grader from an elementary country school attempting to read Einstein. He also recommended that I read Sigmund Freud, Carl Jung, J.J. Rousseau, Voltaire, Darwin, and De Hostos, Unamuno and many others I cannot remember. He even loaned me a copy of Rousseau's Social Contract in Spanish. He advised me not to look for the soul in Rousseau. Instead, he talked with his daughter Maggie, attending a College in another city and loaned me a book in Psychoanalysis by Sigmund Freud. Throwing his arm around me and laughing loudly, he said, "You will find your soul either with Freud or Jung, I promise you." Well, that was a welcome relief, above all, taken from a well-read and highly respectable man. My brother saw me reading Rousseau and said it was a very good book. He wondered where did I get it from and told me that he loves talking with Mr. Clemens. Within a week I was also reading or best, trying to read and understand Sigmund Freud psychoanalytic theory. That was a lot more than I could chew. Mr. Clemens's daughter personally had loaned the book to me saying if I did not understand anything to feel free to discuss it with her. That sounds like more of a threat than a challenge to me. How could I attempt to read college-level books if I was practically a school drop-out? For a moment, I reflected on what she had said to me. If I did not understand anything, I come and discuss it with her. I began to feel relax and less defensive. She was inviting me to discuss not only Freud but the search for my soul. That was how I understood her invitation.

SIGMUND FREUD, MY GOOD OLD FRIEND

I put aside Rousseau and took Freud's challenge. My mind or best, my head was filled with my soul, Freud and Mr. Clements daughter. But above all, I was concentrating on what she had said, "If I do not understand anything… I come to her." What I had in mind was to go to her for my quest in search of my soul. When I had read about the first five pages and I was ready to put the book aside, I asked myself, "How could I dare to read a book in a subject that most people had never heard of it?" However, I became interested in concepts like consciousness and unconsciousness. For some reason that I do not remember, I connected unconsciousness as the locus for the soul and my brain to get together and communicate with each other. In all honesty, I have to accept that I was totally ignorant of both subjects, but mainly, the unconscious processes of my brain. Further perusing the book, I became aware of Freud's emphasis on sexuality as the primary molding factor in a person's life. I began to fantasize whether my soul was in any way connected to Freud's theory of sexuality. I began to make wild connections or associations between unconsciousness, my soul and Freud's theory of sexuality. Freud's emphasis on sexuality was something I had never read or anything like it before. My brothers had helped me in the area of philosophy and religion. However, it was Mr. Clemens who came to my rescue. He called me to his shop and without any introduction he said, "I do not believe in Freud's theories. He takes everything to the extreme. He pays very little attention to culture, religion, tradition and genetic input in a person's life and development. It is too simplistic. He claims that we human beings are driven by an instinctual drive he named it, libido. Freud gives us the impression that we are marionette-like creatures manipulated from birth by a sexual impulse." I had a big bite in my mouth I could not swallow and it was choking me.

Mr. Clemens's comments on Freud were good music to my ears. Freud, according to Mr. Clemens, was justifying the soul disdain for the human body. Humans were no different from any other animal on earth. The soul loath for a human body was understandable; an animal had no place in heavens where the soul came from. The more Mr. Clemens talked about Freud's exaggerated emphasis on sexuality as the instinctual drive behind every man accomplishment, the greatest of my curiosity in the subject. (Mr., Clemens never learned that later in adult life I became a lay psychoanalytic trained psychotherapist.) However, along with Freud's book, Mr. Clemens had loaned me a pamphlet on C.G. Jung. It was as confusing to me as it could be. Besides, the conscious and unconscious, Jung added the collective unconscious. My brother was familiar with Jung's writings through a review from another author. It had a religious connotation and for the moment, I put Jung aside. For reasons that I could not understand, my brother was connected to philosophers Miguel de Unamuno, Ignacio de Loyola and Teresa de Avila with Jung's philosophy. Recently, I came across a book that threw me back to those early years. It says," And in paraphrasing a sentence by Ignatius Loyola, putting it into psychological terminology, Jung states: Man's consciousness was created to the end that it may (1) recognize…its descent from a higher unity…;(2) pay due and careful regard to this source…;(3) execute its commands intelligently and responsibly…; and (4) thereby afford the psyche as a whole the optimum degree of life and development…" [3] In retrospect, there is no doubt that I owe a lot to Mr. Clemens and my brother for their faith in me. My mind was set on the soul as if it were something I could point out and say, "There it is." It is not a small thing. Aristotle, Thomas of Aquinas, Descartes, Loyola, Teresa from Avila and most of the people I came in contact with throughout my life tend to describe the soul as picky and discriminating, just as I saw it as a growing up adolescent.

---

[3] Self and Liberation, the Jung/ Buddhism Dialogue, edited by Meckel et al, Paulist Press, N.Y. 1992, p.276

JUNG'S CONFUSING UNCONSCIOUS

Under my brother, and especially, under Mr. Clemens, I began to appreciate a significant difference between Freud and Jung conscious and unconscious. However, what provoked me to look further in my head, as I used to tell myself, was the Jung complex, extremely confusing but interesting concept of collective unconscious. My first question was: What is it? Does it have any connection with the soul? Is the collective unconscious just an animal attribute? I am sure many more questions popped up in my mind, but these few were enough to keep my mind busy. Along with Mr. Clemens' thinking, I found Freud's work somewhat noxious, relatively easy compared to Jung's exploration into myth, cults, and religious philosophies and, what I had called his obsession with life after death. Nonetheless, I found it amusing and interesting but often contradicting. To justify what I thought was contradicting, I used to tell myself, C.G. Jung is a medical doctor and claims he is a scientist. How come he seems to be dealing with what appears to be witchcraft, superstition, spiritism and primitive religious beliefs to me? I beg you forgive my adolescent hood ignorance. C.G. Jung writes: "I pursue science, not apologetics or philosophy…Only facts concern me… Let us get the facts, for about on these we can all agree… In my view, it is quite perverse to criticize my scientific work, whose aim is to be nothing but scientific, from any standpoint other than that which is appropriate to it, namely the standpoint of scientific method… There is nothing unusual about my scientific method; it proceeds in exactly the same way as comparative anatomy…"[4] I have to confess that I had placed the medical profession on a pedestal; it was truly a pure science-based on long-proven hard facts and laboratory research. But, and this is a big but, both Freud and Jung, as I was reading them, were no better than my grandfather's collection of medicinal herbs and roots.

---

[4] Gerbhard Frei, The Method of Teaching of C.G. Jung in God and the Unconscious, The Harvill Press, London,1952, p235-36.

VIVA ERIC KANDEL, THE NEW STAR ON MEMORY FORMATION

It was my conviction during the time I was reading Jung that in any animal's anatomy, you could see and touch the brain it if you wanted to have a better understanding of it. The soul, on the other hand, now as it was then, meaning during my adolescent hood was something far removed from sight and touch. Further confusing I at present, not to mention during my early youth, Gerbhard Frei add another quote attributed to Jung that says: "As a science of the soul, psychology must confine itself to its subject matter, and avoid overstepping its boundaries by any metaphysical assertion, or by any other expression of belief or opinion."[5] Whether this last quote on Jung is correct or not, it seems to confine psychology to the soul, not to the human body. At the time of Jung's research and writing, he must have read Pavlov, Thorndike, Hall, Skinner and the rising power of behaviorism as the scientific approach to behavior and psychology. Jung was too busy in mythology and the collective unconscious. Of course, in America, Freud Psychoanalysis overshadowed any other approach to psychology until late 1970. Later on, cognitive behavior therapy and Biological Psychiatry took up the upper hand. Most recently, Molecular Biology along with gene sequence research is at the forefront of psychology and brain disease research. Psychology moved on from mind speculative theories to a truly research scientific discipline. Eric R. Kandel, a Nobel Laureate writes: "The underlying precept of the new science of mind is that all mental processes are biological- they all depend on organic molecules and cellular processes that occur literally in our heads."[6] About mental illness, he states: "The

---

[5] Ibid,

[6] In Search of Memory, the emergence of a new science of mind, Eric Kandel, w.w. Norton R. Company, New York, London, 2006, p. 336.

genetic components of these diseases arise from the interaction of several genes with the environment. Each gene exerts a relatively small effect, but together they create genetic predisposition- potential- for a disorder. Most psychiatric disorders are caused by a combination of these genetic predispositions and some environmental factors."[7] The way I understand it now, there is an abyss of insurmountable depth between Kandel scientific research and either Freudian or Jungian research. This is truly a scientific method. At the time, I could not even imagine that Biology, especially, Molecular Biology and neuroscience would lead the way in depth psychology, mental diseases, and disorders. The study of the human mind had been the pass-time of philosophers and religious leaders seem interested in maintaining the traditional thinking and arguments of Aristotle, R. Descartes, St Agustin and many supporters of the dualistic theory of human beings, namely; separate body and soul. I believe it is fair to add that Socrates' supernatural soul was further reinforced by the Christian religion. Whether early Christian found a suitable ally in Greek religious philosophy to justify their own doctrine is still debatable in many quarters. For many of us raised in the Catholic Church, the soul had no other origin, but G-d. I may add that sin and guilt were part of my struggle in search of my self-identity. St. Augustine's two great works as a Christian theologian, Confessions and the City of God were among my brother's favorite books. His teaching must have hit me hard.

---

[7]   Ibid, p.338

St. Augustine on Sin

Later in life, I came across a book loaned to me by a friend. I do not know his motive but in the book, Augustine states that: "The power of sin is such that takes hold of our will, and as long as we are under its sway, we cannot move our will to be rid of it. The most we can accomplish is that struggle between willing and not willing… which shows the powerlessness of our will against itself. The sinner can will nothing but sin."[8] I must have been carrying such a load with me after reading such books that I could hardly pay attention to anything else. You evolve as a person, initially by emulating people closest to you, especially your parents and siblings. You go outside this inner circle and you grow with teachers, men and women whether young or old that by their charisma, love, and care for others, consciously or unconsciously you assimilate it into your own personality. Unfortunately, you continue to fill your brain from teaching of writers and personalities that tend to fit and enlarge, and perhaps improve your already established mental and emotional schemas. Schemas are a point of reference in your brain that will guide your development through life. As an adolescent, no doubt I was struggling with Aristotle's soul, St. Augustine's sin; Freud and Jung that my guiding schemas were in a constant struggle to attain control one over the other. I was unsatisfied and incomplete with what I had inside my brain. An incomplete experience must be provoked or coaxed by an initial schema to burst open for me to continue to grow. That led me to anxiety, healthy anxiety in search of my own identity. Mr. Delgo's toxic memory never left my brain, not even when talking with Maggie. I will talk about Maggie later on. Freud's libido and St. Augustine's over-imposing sin were hitting my brain from all directions. Poor Jose, I pity him now. I do not know if I should blame

---

8   Gonzalez, Justo L. The Story of Christianity, Harper, San Francisco & New York, 1984, p. 214.

Socrates' soul, my mother or mama Tiny. I like to say: Do not blame yourself, reason out a causal factor and keep on moving.

You must travel in mind, time and place, of me coming from a rural setting to a town with limited intellectual resources trying to deal with the prevailing theories and beliefs of body and soul. Adding to my limited formal education was my "obsession" finding a locus for the soul to meet and communicate with the body. Adding to the mental and emotional turmoil of my youth was my jump into Hindu religious belief, Jung and Freud psychoanalysis. Of course, my brother and Mr. Clemens's support must have helped me from having a mental breakdown. Trying to find my own identity either under Freud's psychoanalysis or Jung collective unconscious was more than I could ever comprehend. However, instead of giving up and put the books aside like I felt doing many times, I persevered; I kept on target, waiting for the light at the other end of the tunnel. It seems that I equated the tunnel with Jung unconscious psyche. Later on, I found that Jung himself wrote: "The unconscious is the psychic area with unlimited scope. It is the matrix of all potentialities... is best imagined as a fluid state which has a life of its own..."[9] The same author, further elucidating Jung's theory adds: "At the basis of separate individual consciousness and the unconscious behind it, is the collective unconscious, the common heritage of all mankind and the universal source of all conscious life."

---

[9] Meckel et al., Ibid p.276.

An airplane ticket to New York City

In retrospect, it was a blessing for me at the time not to follow up on Jung. For a reason that I still do not understand. I placed Jung's book aside and concentrated on reading Freud. I cannot say that I understood psychoanalysis as per orthodox Freudian, but I spent a good time wondering and often challenging his "exaggerated sexuality." There was no doubt that I was trying to understand my sexual arousal and interest in girls on Freud's libido theory. At the time, I was courting a very beautiful and religious young lady. I could not accept that my interest in her was sexually dominated. I used to tell myself that our celestial souls were, in a way, responsible for getting us together and not a sexual drive as I understood Freud. How naïve I was raised out in the countryside. Hindu religious philosophy and the Bible are full of Immaculate Conception tales. To my disappointment, two unexpected things happened that changed my life. It took place around the beginning of January 1949. I was 18 years old. Mr. Clemens closed his shop and went to live in the city where his daughter was attending College. My discussion on psychoanalysis was limited to my brother and he was not a fan of Freud. The following month, the girl I was courting left for New York City. She did not write me from New York for reasons it took many years for me to find out. In the meantime, I found the address of a friend of mine from childhood. I wrote down his address and place it in my wallet. During the first week of April, I found myself flying to New York. When I arrived in New York City I pulled out my friend's address and asked a taxi driver to help me out. To my surprise, the address I was given was located in Philadelphia. I still did not know how far one city from the other was. The taxi driver offered to take me to a public bus station that took me to Philadelphia. Luckily, I did not have problems locating my friend and his family.

My home in Philadelphia

My friend was able to get me a job at a local newspaper cafeteria. I rented a single room in a private home of a Roman Catholic German family. By coincidence, the landlady's brother Paul had volunteered during the Spanish civil war in favor of the Republic. He spoke fluent Spanish. The land lady's daughter Maggie, around age 26, took the like of me and facilitated me learning English. She asked me for the names of American authors I would like to read. When I mentioned Freud and Jung she took a look at me that made me feel embarrassed. I thought I had said something bad. She asked me to write down their names. The problem was that my pronunciation of those two names was far removed from the correct one. We went to the local public library and she made it possible for me to borrow books. I do not remember the title of the book that related to Freud, but she was very helpful to me. She clarified to me many things on Freud that I could not understand by myself. Her focus was the neurotic behavior we experience as a result of the struggle between ego, super-ego and impulses. Meanwhile, her uncle used to take me to Independence Park at the center of the city to listen to his friend's arguments in favor of Socialism and the evil of Capitalism. Of course, I preferred the young lady's company, but I was not assertive enough to choose. Maggie used to work in a Bank and attended College during the evening. Occasionally, she used to bring home books for me to read and then discuss it with her. Thomas Jefferson and Benjamin Franklin were two of my favorites as well as the French philosopher Voltaire and Rousseau. I became deeply interested in the American and French revolution. Mr. Clemens had talked to me about the wonders of Paris and Maggie reinforced my fantasy in the city of enlightening wisdom.

A BITTER/SWEET MEMORY, JOSE IS DRAFTED INTO THE U.S. ARMY

My happy and fruitful first two years in America came to an end too soon. In 1951, I was drafted in the Army and destined to Germany. My first three months of basic training in New Jersey were the worst nightmare I could ever experience in life. My platoon Sergeant was from New York City and he must have had a rough time with Puerto Ricans in that he projected and discharged all his anger and hate towards me. That was discrimination in big. My toxic memories about him haunted me for so many years. I felt defenseless, incapable of saying or doing anything to defend myself from such an unjustified abuse. His behavior was many times worse than my school teacher, Mr. Delgo. Twice I visited Philadelphia, but I was afraid to tell Maggie and her uncle how I was punished and discriminated against by the Sergeant. After I completed my basic training in New Jersey, I was sent to Germany. While in Germany, I visited Nuremberg where the Allies held the tribunal for Nazi criminal officers. Later on, I visited Frankfort and Munich, but the city in Europe that had always fascinated me was Paris. During my first visit to Paris, I could not do anything but look in awe and with admiration for such beauty. I ran to see and go up the Eiffel Tower, the Notre Dame Cathedral, Louvers Museum and the Bastille, place of reference for a Tale of two Cities. From the top of the Eiffel Tower, I looked around the great city Mr. Clemens talked to me about in disbelief. As if I were looking at a movie, the French revolution of 1789 with its King, Marie Antoinette, the guillotine and all the violence that it produced rolled on my mind without stopping. I just walked along the city's great Boulevards just to tell myself, "Yes, I am in Paris, the center of the world's intellectual life." What a difference from my early childhood experience around my oldest brother's grocery store. To appease my ever curious and non-conforming head, I used to tell myself, "You have had difficult and unfair treatment by some individuals, but in the end, it paid off." Next in line during my soul searching was my trip to Spain in 1952. My mother was always talking about the wonders of the country and the accomplishments of its people. When in Madrid, I did not know where to go. There were so many things I had read about the country, that when I walked La Castellana Boulevard and La Gran via in Madrid, I could not do anything but sit down and look around in disbelief. My first visit to the Royal Palace and all its beauty was like walking in a fantasy world. Later on, when I went to the Prado Museum, and me in person, looking at Goya`s Majas as well as other great Spanish Masters, was as if I were daydreaming. Out of joy, I used to pinch myself and say, Yes, Jose, it is you; you are now watching and appreciating some of the best paintings the world has produced. I pinched myself to be certain that it was not a dream; I was not trying to fly like birds nor walk on stilts to avoid the dirt around me.

JOSÉ GOES TO COLLEGE, OLD SWEET MEMORY

After my trip to Paris and Spain, my search for the soul and its communication locus with the body took another turn. I began to look for the source of my anger towards Mr. Delgo and the platoon Sergeant because they were no longer near me and capable of hurting me. I looked for people I could talk with regarding my concern, but either I did not have money to pay a professional counselor nor was I in contact with friends that could discuss the subject with me. I had left the United States for Puerto Rico. After getting a High School Diploma sponsored by the U.S. Army G.E.D. test, I was admitted to the University of Puerto Rico in Rio Piedras. At the University's general library, I found all the books on Freud and other psychologists that I needed to consult. I had learned that traditional psychoanalytic psychotherapy was aimed to uncovering the unconscious and making it conscious. I had placed the soul aside, at least, momentarily and went back to Freud. As I have already said, Jung's theory of the unconscious, which I found different from that of Freud's, was too complex for me to understand at the time. Despite Freud's exaggerated emphasis on sexuality, his approach to the conscious and unconscious memories seems more appropriate for me to deal with my own conscious and unconscious memories from childhood. It was hard for me to accept that I had an emotional problem on top of Mr. Delgo's abusive verbal behavior and physical punishment. I do not remember where I had read one of Freud's books that said, "No neurosis is possible with a normal vita sexualis." I had diagnosed myself already following Freud's criteria. I was wondering where during my childhood and adolescent's sexuality (libido) had intervened to make me feel neurotic. At the time, I was studying in College, An Introduction to Psychology, and consulted a professor who referred me to a counselor. Well, neither Freud nor Jung had made an impression on the counselor and he began lecturing me about growing up in the country and city life. I never made it to the third session. Luckily, I found a third-year student of Sociology, named Tony, who was interested in Freud's work. At least twice a week, we got together to discuss our behavior. Most of our conversation dealt with childhood memories. He had confrontational problems with a teacher during high school years. We used to call the teacher's names and curse them as much as we could. We felt relief just venting out on our feelings to each other. Sometimes we used to carry the sessions outside the library to the open campus. When verbalizing our frustration and anger we began kicking the shrubs around us as if it were the teachers in flesh and blood. We found out more relief to our anger when we took a stick and hit a tree trunk and wrote the name of the teacher around the tree trunk. We kept it a secret between us. We did not want anybody to find out the motive of our game. As if it were a childhood game, my friend came out with an excellent idea; he drew the picture of my teacher and his teacher on a paper. Next to the cafeteria, there was the recreational area where we taped the teachers, he had drawn on a piece of board and began throwing dots at it. A couple of students began to laugh about it but when we explained our game of mental and physical exercise, they joined and helped us destroy our mental monsters.

AN INTROJECTED MEMORY
IS NOT ASSIMILATION

I compare introjections with something you swallow whole and did not chew. If it is a food you have swallowed, in all probability, you will get indigestion. We usually take a laxative to expel the obnoxious and sickening food out our digesting system through the rear end of the food containing elastic pipe. If it is mental introjections like Delgo's memory and of the drill Sergeant, we must clarify for you that the objects I had introjected were an all-powerful psychic reality. Delgo and all his wicked thoughts and actions were very real inside my mind-brain. Those memories had occupied a space inside the flesh and chemistry of my brain tissue. By the power of his actions and authority society had attributed him, Delgo appears to be a giant of immense strength and unparallel skills to humiliate and torture individuals like me. The image we had created of him was as many times his real size and strength. There is a great difference between the object that abuses you and the mental or psychic reality you are forced to make. There was the torturing object I was carrying with me besides the structural brain damage and changes he had done to me. To better understand my plight, you have to learn to see the difference between the flesh and bones of Delgo's body with the ogre he had created in my brain's neurons. On one hand, you carry with you the psyche reality of a mentally torturing imagined person incrusted in your brain tissue as a monster created by Delgo's grotesque and evil behavior, while on the other hand; you carry an individual, the actual man, a teacher praised by a poorly informed community, including your own parents. When you carry with you an episodic toxic memory, it is not an illusion or a make-believe thought. You have to understand that it exists in mind energy and in a brain made of flesh. It does not go away with a soft hand pat on your shoulder, or by swallowing a pain killer. It may provide brief and temporary relief.

In this case, your organism has a permanent toxic producing tenant inside your most developed, exciting and marvelous group of cells, your brain. This tenant does not take vacations or goes to the theater unless you take it with you. This mental and physical bug stuck at the synaptic level, would be the source of depression, anxiety, insomnia, headaches, panic, and many other mental disorders.

TONY'S HELPING HAND

What my friend Tony and I did not realize at the moment was that we were doing something novel in psychology. We were not only bringing out to the open toxic memories, but we were also putting actions and feelings into the unconscious. When we wrote the name of our teachers, the hated objects; we were in effect killing or destroying an old introjected (an unconscious process) object. The teachers' names, face, body, and mood were very clear in each of us, but what we did not realize at the time were our own repressed aggression and inner strength. We were destroying our past bad memories and replacing it with new synaptic connections. Our words when we were challenging the teachers were not empty words, but an emotionally charged, action propelled and consciously directed brain action. It was an emotional organism in action, not an abstract mental exercise. And, it was not the soul directing the body into action. It was my brain in control of my body that was engaged in eliminating from my mind-brain - a bad and abusive object. The brain-mind and physical body were acting in a coordinated fashion to eliminate a threatening object, a memory. Tony and I clarified, taught and helped each other identify our problems. We pointed to each other our limitations as well as our strength. He was very helpful making me understand the influence of culture and organized religion in the shaping of our personality. We were not only destroying bad memories; we were making use of our own resources. The emotional arousal we developed while jumping on Delgo involved all parts of our body. By listening to myself repeat my strength as seen by Tony, I was re-discovering assets, Delgo had taken away from me. My new thoughts and memories about myself were beginning to change my behavior. The truth about myself was coming out of darkness; my brain was set free to learn and discover the universe around me. I felt that I was getting close to, "The truth shall set you free" My friend Tony and I were not engaged in any storytelling of any sort. We were not repeating a story for entertainment. We were addressing the most fearful and socially taboo feelings and thoughts each person dares not to share with anyone else. We had agreed upon ourselves not to get angry at each other for pointing out insights and interpretations we experienced about each other. As I later learned during my sessions on a couch, I could associate all I could, but the emotions behind it remained locked in the box. We are an emotional organism. Like all animals, the brain stem and limbic system (emotions) were first to develop long before the frontal cortex came to play its executive role. Although the brain developed in stages, it works in a unitary fashion. The prefrontal cortex and the limbic system coordinate its actions. The amygdale, very often described as the seat of fear and emotions is a very primitive nucleus. It is located in the frontal medial temporal lobe of each hemisphere. Because it responds to fear, when absolutely necessary, it functions outside the control of the neo-cortex and has saved our lives in the past and the present.

KAREN HORNEY SELF-ANALYSIS, A HEALTHY MEMORY

Among the books or quotations that I read while trying to understand my repressed feelings, I came across Karen Horney which reads, "Feeling helpless means a weakening of the very foundation of self –confidence. It carries the germ for a potential conflict, the desire to rely on others and the impossibility to do so because of deep distrust and hostility toward then."[10] Sometimes I wondered if my distrust in teachers and their substitute or nominee ever prevented me from following my psychology professor's recommendation to the counselor. Very often, through identification, unconsciously, we project our feelings towards persons that may have traits and behavior similar to the abusive person. Unfortunately, persons closest to us like our parents or their substitute as well as our teachers in primary grades abuse children leaving long-lasting emotional scars. The abuse does not necessarily have to be physical punishment. Emotional neglect can prevent or delay the child's physical growth, learning and brain development.

My friend and I, although we did not have the slightest idea of what was happening in our brains, we were building new synapses. New molecules were synthesized for new neuronal circuits and pathways. My brain was getting richer and I was getting smarter. Many researchers have proven in laboratory trials that it is true; new synapses are established and learning takes place. My friend and I were not engaged in an abstract game trying to get rid of our torturing memories; we mobilized all our brain and body into coordinated action. During the process of destroying our pain-producing image of our teachers, we were experiencing great pleasure. It means that

---

[10] Horney, Karen. The Neurotic Personality of our Time. New York, W.W. Norton,1937, p. 82

dopamine-producing brain cells were having a good time making multiple connections and exciting other cells. Dopamine is often called the pleasure producing neurotransmitter in the brain. It often takes hold of you and you become an addict. It is also involved in at least two frightening brain diseases, namely schizophrenia, and Parkinson's disease. Along with dopamine, we have endorphins as part of our pleasure system. Endorphin is short for internal morphine.

MEMORIES OF SELF-CONFIDENCE AND SELF-EXPRESSION

As dopamine was making me feel good by weakening the toxic object, the teacher; my fear of anyone resembling him was fading away. It meant that I could express my opinions freely in the classroom. When I needed a comment on something the professors had said and I did not feel comfortable with it, I said it without fear of retaliation from my introjections. (I recall when I was supervising psychotherapists that I used to compare introjections with swallowing whole, a mule with its tale and hooves.) As I was speaking out my feelings in and outside the classroom, I began to notice that I was having self-confident and gaining the respect of my classmates. I did not have to go home after each class and ruminate that I should have said this or that to the professor. I must add that psychology books at the time were centered on Williams James's introspection, John Dewey and Wilhelm Wundt. There was Pavlov's classical conditioning as well as John B. Watson's behaviorism. James, Dewey and Wundt psychology never attracted me. Pavlov was interesting but I did not know how to use it. On behaviorism, I wish I could have paid more attention to it. However, coming from a strong religious background, Darwin and Watson's animal approach to human ('s) existence and behavior was, perhaps, too threatening for me. Besides, I was after the soul and the unconscious, primarily, Freudian unconscious. However, feelings and action-oriented approach to destroying toxic objects have more in common with behaviorism than Freud's theories. I was actively challenging and sensitizing myself to the fearful and painful object. When at home I made a man like the figure of sand and other material that fit my objective, placed my teacher's name on it and whipped him until it broke to pieces. We want to emphasize the role of emotions involved during this practical exercise. Mr. Delgo was castrating me, a psychoanalyst would say, but my brain in full action was taking that power away from him. I was dancing in joy on his face and his genitals. The psychic reality that I was forced to accept was weakening fast. It was losing its grip on my brain. I was not a passive observer and recipient of a toxic memory. My brain was developing very useful tools to defend itself from the inside out. It is not new; your brain has done it too. Remember, your brain is always working for you. It sends you signal you must learn to interpret and use for your own benefit. Whether the following quotation from Jung applied to me at the time is irrelevant or not but, it seems quite appropriate. It states: "The personality of the depressed individual is made up of early introjects which prevent him from contacting his true identity…; he is crushed by the contents of the unconscious. His self-confidence dwindles…"[11] I was listening to signals from my brain and putting it to work.

---

[11] Jung, Carl G. Two Essays on Analytical Psychology, New York: Pantheon Books, 1953, p. 136.

WHEN DELGO INTROJECTED MEMORY HAD THE UPPER HAND

My ever toxic introject with Mr. Delgo, did not only prevented me from self-contact; he was destroying me physically, mentally and socially. It was preventing my brain from making appropriate and healthy connections. I could not express myself freely; my creativity was dwarf by the intrusion of a torturing memory that even during sleep did not leave me alone. The content of my dreams were monsters haunting me even in the house of prayers. My understanding at the time was that if I destroy the toxic memory of Mr. Delgo, I could bring together my conscious and unconscious mental content then, it would facilitate my own growth. I thought Jung had addressed this issue when he stated: "Individuation means becoming a single, homogeneous being... it also implies becoming one's own self...coming to selfhood or self-realization... means a process of psychological development that fulfills the individual qualities...man becomes the definite, unique being he in fact is."[12] I wanted to become an individual free of past torturing memories that prevented my self-realization. Although Jung remained a difficult person for me to understand (,) selectively, I picked what I thought would help me the most. For a reason that I still cannot explain, I have always believed that my organism is equipped with innate tools for self-healing. My earliest observation in this theme was watching how a wound in my arms or legs heal itself with little care by me or anyone else. I had always wondered how and when did my organism was equipped with a capacity for self-healing. It was not until I entered college that I learned about our immune system, B. and

---

12 C.G. Jung, Two Essays on Psychoanalytic Psychology, Pantheon Books, New York,1953, p. 171-2

T. cells and the body's potential to regenerate cell. Growing up in the countryside helped me watch animals recover from wounds inflicted by other animals. The power of self-healing in animals was always a curiosity while growing up.

LEARNING MEMORIES OF BRENNER AND FREUD

At the time, there was a small book by Brenner that stated, "The two fundamental hypotheses of psychoanalytic theory are the principle of psychic determinism and the proposition that consciousness is an exceptional rather than a regular attribute of mental processes. Psychic determination established that nothing happens in the mind by chance. Each mental event is determined by one which preceded it. Conscious mental life is a small fraction of all psychic activity. Mental development is based primarily in the oral, anal and phallic stages with successful resolution of the Oedipus complex. Sexual energy plays a determinant role in the formation and development of human personality and behavior."[13] It is a mouthful of information not only difficult to understand, but it looked to me as toxic as Mr. Delgo. I carried it with me for many years, but despite its appeal to me, it remained stuck in my throat. I was unable to swallow it or to throw it up. That consciousness is an exceptional rather than a regular attribute of mental processes is a scientifically proven fact. Modern research and technology leaves no doubt about it. Also, there is no doubt that mental life is a small fraction of all psychic activity and is another truth of psychoanalysis. It is what happens inside our brain that is an infinite complex and unknown is equally true. Freud opened the brain to all sorts of investigation. Freud, along with Cajal and many others thereafter, opened the doors to investigate our most sophisticated organ in our body. However, when Freud added to his theory that mental development is based primarily in the oral, anal and phallic stages, it provoked on my goose pimples. As far as I was concerned, he had left out most of the human being and chose to

---

[13] Brenner, Charles, an Elementary Textbook of Psychoanalysis, New York: Doubleday Anchor Books, 1955.

deal with the beast. I understood that he meant that king libido (sex and aggression) was responsible for us to build houses, beautiful palaces, develop art and cultivate beauty in all its forms. I struggled for many years between Freud and my soul. Freud was on the side of the human body and the soul running away from it. The chemical soup that was able to achieve life wants survival at all cost. As long as it lives, it will live anywhere and in any form regardless how it is achieved. Although there are ample, well organized and logical arguments in its favor, man has not destroyed himself yet, and continues to be a creative creature capable of harnessing his libido. Furthermore, we can argue that a libido dominated beast like Hitler, Stalin, Napoleon, Attila, Alexander from Macedonia and power-hungry individuals are an exception in human history.

A HUGE ICEBERG, THE UNCONSCIOUS CONTENT OF THE BRAIN

In retrospect, how could I have pretended to find the locus of communication between my soul and my body, especially on my brain? Freud, Jung and as well as the modern technology have demonstrated the depth and complexity of the unconsciousness of the human brain. Molecular biology and brain science in general have made gigantic advances unlocking and revealing the secrets of the brain we have just scratched the surface of the iceberg. Many authors have compared the unconscious to an iceberg like the one that destroyed the Titanic in its maiden voyage to America. We can only see the iceberg that is above the water, but the huge part beneath it remains unknown. Many times in the past I equated Jung's personal unconscious with the iceberg above the water and the collective unconscious with the iceberg underwater. It refers to size or volume and not to content. When I was dealing with memories of Mr. Delgo on conscious memories from my childhood, the flow of emotions that came out went deeper into my personal unconscious. Although I was not aware at the time, I was also questioning my father's role while I was growing up. While kicking and destroying Mr. Delgo's make-believe body (a dummy) at home, these flashes of my father's occasional angry mood took front stage. Mr. Delgo was considered one of the best teachers the school ever had. My father considered him a very good and a disciplinarian teacher for the good of the community. I never told my father of Mr. Delgo's abusive behavior towards me. Later during therapy, I realized that I was blaming my father not only for not coming to my rescue but also for not looking beyond the surface on an object, into the deeper layers of anyone's personality. It means that I had over-evaluated my father. He was considered a very respectful person by most families in our neighborhood. He had to stand up and above most individuals (;) that was what I was expecting from him. It was in my mind but never verbalized. I began to blame my father along with Mr. Delgo, but in a different manner.

JOSÉ LEAVES COLLEGE FOR NEW YORK CITY

When I was two and half years in College, I had to leave because I ran out of money. I was studying under a program sponsored by the Veterans Administration. I migrated to New York City and held two jobs. I began working at 8:00 a.m. in a factory as an assistance shipping clerk and at 5:00 p.m. I was working two blocks away in a restaurant until midnight. I could only sleep four to four and half hours a day. I had the weekend to make up for it. I worked for about eleven months and saved enough money to finish my College studies. During the last two years in College I became attracted to Independence for Puerto Rico movement known as F.U.P.I. This group was composed of College students. We were an aggressive, but non-violent group. We felt that the power of persuasion plus mass public demonstration would persuade the American government to grant Puerto Rico its independence. Personally, I had great admiration and respect for Lolita Lebrón and the group that took upon them to show their frustration by shooting some Congressman in the U.S. Capitol. I have had a love and hate relationship with United States. I love its people and institutions, but hate its policy towards Latin America. At times I identified with the Southern cause, but hated slavery and discrimination based on color and race. Strange enough, it seems tied up with my father's feelings and attitude. My father used to take pride in his family's military history in the Spanish Army. But he was equally proud telling us that he was ready to take arms against the American invasion of Puerto Rico. Around forty years after the invasion, he used to show us the rifle he had used to join the Puerto Rican and Spanish militia against the invading forces. However, he had great respect for the American system of government and institutions. It was my mother in 1945 that threw my father's rifle, and my brother's pistol upon his return from Germany, into a latrine.

GABRIEL, MY GRANDFATHER
AND GREAT BOTANIST

Besides my father's proud history, I still have vague memories from mother's father. He was the first member of my mother's family to be born in Puerto Rico. He was a proud Spaniard whose loyalty to his mother country never gave way. I was five years old when he died. He was a self-made botanist and healer. During the evening, mostly on weekends, the whole family used to get together for him to tell us stories about themselves that still resonate in my head. I believe that it has to do with my binary relationship and feelings towards United States. I wanted to go on with life and leave behind my torturing memories. Of course, this was ignorance and wishful thinking on my part. However, I was always reading and I must have had a strong conviction that I could heal myself. Freud, Jung and other philosophers and psychologist must have taught me to persevere in my endeavor. From Heinz Hartman I had learned a lot but recently I found another psychologist that best prescribes, fits to my belief. "We know that nature does not operate by design and does not decide in the way artists and engineers do but this image gets the point across. All living organism from the humble amoeba to the human are born with devices designed to solve automatically, no proper reasoning required, the basic problems of life."[14] I must have acted in ignorance but there was always my "device" that guided me in the right direction. I never said it was going to be easy but 90 years of struggle have not been in vain. When my friend Tony and I in College began to kick Mr. Delgo, my hated introject, we probably did not really know exactly what or why we were doing it but we felt relieved, stronger and persevered until today. Soon, I came to the realization that it was not only my teacher that I had to confront and attempt to resolve my

---

[14] Antonio Damasio, Looking for Spinoza, Harvest Book, New York, 2003, p 30.

unfinished business. I was equipped with a brain capable of taking care of itself. I had to learn to listen carefully to signals coming inside myself. I learned that my organism had survived many challenges for thousands of years. When it fitted my needs, I used to call on my soul but at the same time, I used to say to myself, keep on track and you will be compensated. I became aware that while consciously facing and actively challenging Mr. Delgo, I had uncover a huge load of emotions covering my entire family as well as friends and the community in general. I began to think about my siblings especially my elder brother and my mother. Neuroscientist Damasio states, "For an emotion to occur, the site must cause subsequent activity in other sites... As with any other form of complex behavior, emotion result from the concerted participation of several sites within the brain system."[15] Dr. Damasio is a neuroscientist and an author of very interesting books that I enjoyed reading. His observation comes from laboratory work. My confrontation with Delgo was a real life experience. No doubt my memories covering other family members came from several sites in my brain.

Dr. Damasio tells me that my torturing memory of Mr. Delgo had invaded many parts or sites in my organism but primarily, my brain. The brain is the supreme commander of our organism. There was not a single part of my body that was not touched and harmed by the poisonous image of Mr. Delgo. It was interesting that I was not just talking about him; I had my whole body participating in destroying his memory and restoring my strength and identity. The emotion of Mr. Delgo's abusive behavior arose in my brain included many sites. Among those sites must have been the amygdale, hippocampus, prefrontal cortex, hypothalamus, thalamus, the limbic system and many others. I could say that my whole body was, in some way, hurt by his memory. For many years, I would feel my heart racing and blood pressure go up at the site of a person that looked like or spoke like him. It is important to point out here that while I was doing therapy I came across many individuals that their life happiness was cut short or destroyed by toxic memories like Delgo.

---

[15] Ibid, p. 59.

Memories of this sort have no boundaries, religious beliefs or social status, unfortunately, it is taking place right now not only in schools, but in the church. And the victim remains isolated and in the pain. The perpetrator is a powerful figure in our society. The abused child is ignored, often punished for revealing the action of a wicked person.

FORMING A TEACHERS
GROUP FOR THERAPY

Before we go back to our group, I like to say that after graduating from College in 1960, I went to teach history in the 10th, 11th and 12th grade. Around 70% of the students went on to College. It has been the most fruitful and gratifying teaching experience I ever had. My students were very intelligent and motivated. They also motivated me to go on and discover my true identity. Consequently, another teacher and I organized a group for learning and self-growth. I shared with him my learning experience with Tony, my sociologist student in College. The first two meetings were wasted addressing multiple topics were not interested in or were not of concern for the whole group. Among the subjects we brought up during our first two sessions, was students' learning and discipline in the classroom. However, we turned it down arguing that it was for the school Principal and Supervisor's responsibility in addressing such a complex problem. There were good justifications for the argument that the school administration, parents and community in general had to be engaged to assist those issues. There were two lost meetings discussing something that I felt was out of our competence and concern at the moment. The truth, I was interested in talking about ourselves as teachers. I wanted to leave the school out and take each of us as the subject of discussion.

I suggested that a student's problem might will be a teacher's problem. I said that a teacher might be transferring his personal or family problem to the students. There was silence in the group. After a short while, a group member said, "Lets us find a subject we all can agree to." I sensed that the group was afraid of self-revealing sessions; they were behaving defensively. I volunteered to be the first person to begin our self-search. I recall I said that I would be very happy if they could help me solve my problems. The balls of their eyes pop up like a car lights at night. They were puzzled. It was a serious challenge. I had confessed that I had problems and want them to help me. I was not only challenging them as to how much they knew about human behavior; I was challenging them to be brave and honest enough to be able to open their fears in the group when the time comes. My closest friend, Rocky said, "Jose, you have been trying to kill Mr. Delgo ever since you were in College. Now it is your chance, we will help you finish the job." Incidentally, two members of the group had Mr. Delgo as their teacher, too. Immediately, the level of anxiety came down and we could see smiles on their faces. The fear of self-revealing passed from each teacher to myself, an outside object willing to reveal his fears and sins. Of course, I was anxious. My heart was giving me messages, my pulse went up and my head flow of blood was reaching every cell in my brain. I do not remember if I asked my soul not to leave me alone but I clearly remember saying to myself, "on target, Jose, on target!" In a way, I was repeating my father's old prayer.

THE FIRST REAL GROUP MEETING

I began telling the group how, where and when I had started my boxing match with Mr. Delgo. I had shared with Rocky part of the story and always anxious to know more about it. He immediately supported me and volunteered that he wanted to be the second after me. I said it would be many more sessions before he could take over. At the end of the first meeting, I sensed that everyone in the group was very enthusiastic and looked forward to our next meeting. Before we separated as a group, one female teacher said in a very convincing voice, "What Jose brought up in the group, stays in the group. It is not to be shared with anyone else." It received unanimous approval from the group. The group consisted of eight teachers. During the next meeting, to my surprise, a male member of the group, opened his car trunk and pulled out a medium size plastic figurine with the name Mr. Delgo pasted on it. There was an initial laughter by the group. Someone in the group said, "Jose is going to get all the help he needs in here. Then added, Jose, I hope you do not mind." He pulled out a long magic marker and wrote, SATAN on Mr. Delgo's chest. I was very attentive to the group's behavior and noticed two ladies covering their mouths. The group was getting cohesiveness, it was gluing together. I felt relaxed by now. I was the center of attraction just like in my childhood days. When I told the group Mr. Delgo used to make me kneel down on a sand wooden box in front of the class, some members raised their hands over their heads. Mr. Delgo had brought to the classroom sand from the beach and kept it in a box to punish us. I added that I preferred Mr. Delgo to give me ten slashes and not twenty minutes on a sand box in front of everybody. I was very skinny and my knees seemed to be bones only. At the time boys had shorts, not long pants. I told the group that occasionally, I wish I were dead or very sick to avoid Mr. Delgo. At the time, a group member got up his seat and said to me, "Jose, I know your father very well. Had you told your father, Mr. Delgo would have been gone" A lady teacher came and hugged me. It was the first time anyone had shown compassion, for the little child inside me. They asked me how they could help me at the moment. I could not answer them but the same teacher who had written Satan on the mannequin or dummy, slapped it with the back of his hand. Emotions were running high. Empathy and identification with the underdog were cementing.

TURNING ON PAIN BUTTONS

The lady teacher had turned on an emotion site deep in my brain. When I verbalized that on more than one occasion, I wished I was dead not to come to school. I must have turned on everybody else's pain points in their brain. Every group member was trying to help the little Jose, not Jose the teacher. Some emotions were repressed and were attempting to come out to the surface. Repression is an unconscious mental process in psychoanalytic theory. I had revealed myself in a group for the first time. Perhaps I did not understand the full meaning of what I was doing, but I kept telling myself: keep on target, you can do it. There was a moment when my body was visibly shaking and my throat was stuck with something I could neither swallow nor spit out. I tried to say something, but I could not. There was no doubt in my mind that the little child in me was looking for support and love. The little child in me had awakened and touched everybody else's past emotions that remained incomplete and unfinished. There were emotional vacuums to be filled. When one emotion is unlocked, it must be done carefully and skillfully. As Damasio said earlier, many emotional sites were turn on and when improperly done, may do more harm than good. But I was determined to go ahead and luckily, I had a good support group. It was not only Mr. Delgo`s memories that I was struggling with, but it also went deeper.

GROUP RE-AFFIRMATION

After a couple of more intense sessions with the group, I was advised by the group that I should take a recess and try to put in proper perspective all that I learned during the meetings. During the following session, the group discussed briefly how my experience had touched everyone there present. They reaffirmed the principle that everything must stay within the group. My friend Rocky had initially said that he wanted to be the next person to reveal himself. However, the teacher that had slapped the figurine or dummy convinced Rocky to let him be next. For the sake of reference let's call him, Jim. He was a practical person that once told me, "I am like St. Thomas, and I believe what I see." I was reading Socrates, Aristotle, St. Agustin, Descartes, Freud, and Jung. He showed me his Bible as he put it, John B. Watson Behaviorism. I had been referred to Behaviorism several times, but I could not come to see my behavior as postulated by Watson. Somewhere I had heard or read the following: "Never deny in darkness what the soul has given you in daylight." I had read about Pavlov's classical conditioning but I did not know how to use it or simply did not want to let go off the soul. It had to be neither. Jim gave me selective page copies perhaps to prove his point. He underlined the following: "These concepts (religious concepts) these heritages of a timid past- have made the emergence and growth of scientific psychology extremely difficult. One example of such a religious concept is that every individual has a soul that is separate and distinct from the body. This soul is part of a supreme being. This ancient view led to the philosophy platform called dualism."[16] I know he did not mean to offend me or to push his point of reference upon my throat, but at the time I was not ready to embrace the emergence of new psychology. I am in debt with the group forever. The following year, when the group was at its height, I was sent to

---

[16] James B. Watson, Behaviorism, WWW Norton & Co. New York, 1958, p. 2-3.

New York by the Puerto Rican Education Department. A new chapter in my life had just begun. From then on, my life was dedicated to working with children and adults with mental disorders.

WORKING WITH INSTITUTIONALIZED YOUNGSTERS IS A LASTING EDUCATIONAL MEMORY

In New York City I was hired to work with institutionalized children. They were boys under the age of 18 with emotional, learning or behavioral problems. Some of them came from broken homes, but there were children from healthy and well-established families. Most of them came from New York City, but there were boys from well-established suburbs. My supervisor was a trained Freudian psychoanalyst and so was the vast majority of the institution staff. I fell quite at home talking about Freud and psychoanalysis. However, in dealing with the youngsters' problems, I felt uneasy and incompetent attributing their problems to strictly Freudian theory. Among psychoanalyst theorist, some blame the mother for almost everything. They claimed that lack of maternal love and care was responsible for their children's emotional and learning problems. While this might be true in many cases, to generalize it is wrong. I used to visit their homes and interview as many individuals in the household as I could. No matter how hard I tried, Freud's theory did not fit well with the problems I was facing with our boys. I used to argue with another minority staff member that poverty, discrimination, and unqualified teachers (,) in many cases, were responsible for our problems. However, my friend from Venezuela had been working at the institution for over five years and he advised me to be careful. He said my supervisor was strictly Freudian and his recommendation would determine if I go or stay in the agency. That was more than enough justification for me to stay with Freud. I used to maintain weekly telephone or written communication with the group of teachers back in Puerto Rico. They used to inform me how much they enjoyed and benefited from it while learning about themselves. On one occasion, Jim mailed me the following pages of a book for me to read. It began in the following fashion, "The development of an individual is brought about by the interaction of these thousand substances-genes- their interaction with each other, with other parts of the cell, and with material taken from the outside...; the way a given individual develops, what he becomes... depends... other things being equal, on the set of these substances he starts with."[17]

[17] John B. Watson, Behaviorism, op. cit, p.50

MEMORIES FROM JIMS' QUOTATIONS (.)

Jim was a brilliant person and I appreciated his attempt to get me in line with a scientific approach to human behavior. I believe I was justifying a door to Behaviorism and wondered what set of substances did I started with. I traveled in mind to my early childhood but could hardly bring my mother's image into the picture. I could recall my father clearly, but my mother's memories were missing. I went on to say to myself; genes from my father and genes from my mother. I could picture my father from childhood, but my mother did not come out until I was around seven years old. Another part of the quotation that caught my attention is, "material taken from the outside." Did outside mean from outside my body? Did it refer to my soul? In retrospect, it seems I was quite at a loss. I had a job where Freud and his psychoanalysis were supreme, it was necessary for my survival. I was strongly questioning its clinical application in my work setting. Behaviorism was getting closer and closer to me. I could not get rid of Jung because with him I felt I could carry the soul with me. The following quotation from Jim threw me off balance. "Anyone who declines the labor of becoming familiar with the fundamental features of the genetic system and its method of operation cuts himself off from the possibility of understanding the nature of man."[18] Well, adaptation I believe has helped me survive all these years. Perhaps, I have learned from my father. He used to say, "Swimming against the flow of the river is not only hard (,) but dangerous." I was carrying my father's genes. They made him strong, intelligent and proud of himself. He never stopped telling me about his family tree till his great-grandfather.

---

[18] Ibid, p.52

IMAGINAL EXPOSURE

I had begun pedaling (kicking) Mr. Delgo`s dummy and had engaged teachers do likewise with their own introjections. Between Tony, me and the teachers had developed a fairly well-constructed treatment modality for getting rid of toxic memories. Somewhat modified to meet each situation and place, it has worked fairly well for many people. Many years after, I got hold of a book that I believe validated my initial confrontation with Mr. Delgo. It reads (:) Masserman ... reduced an animal's fear and abnormal behavior through conditioning techniques. Wolpe consolidated and expanded these findings and concluded that the most effective way... was to re-introduce the frightened (person) to the fear- conditioning stimulus... provided a foundation of systematic desensitization...; direct exposure was replaced by imaginal exposures."[19] That was exactly how I had begun my own therapy during my last two years in College. With the support of my friend Tony, I imagined Mr. Delgo, the fear object, lying on the ground and me kicking and pissing on him. My friend did not only help me kick him (,) but applauded me after each confrontation. We were practicing action and goal-oriented healing or modern psychotherapy. The teachers' group was also practicing high-quality group therapy. Our body, mind, and soul, if you please, were engaged in a present, life experience consciously getting rid of a very painful individual in our life experience. We were not just passive actors verbalizing what may appear as anger; Mr. Delgo`s dummy was a mental reality. Delgo`s dummy on the ground was real for my brain. The teachers and I improved our initial confrontation with Delgo. Besides using our boots for kicking, we placed tape on his mouth, twisted his nose, wrote the same words he used against me and pasted it around his head. At the end of each session, we

---

[19] Cognitive Behavior Therapy, edited by Clerk & Fairburn, Oxford University Press, 1997, p. 8.

decided not to touch him, but drag his dummy by his feet to a car trunk. As Jim's suggestion, we decided to take the teacher title away from him and replace it with Canis Delgo. The dummy was not torn apart but mended when needed. The group was quite inventive and each one built its own device to punish Canis Delgo for each member's toxic memory. When the anger was directed towards another person besides Delgo, we added the name to the list of Canis Delgo. There is no doubt that brain's physical changes like new dendrites and spines were taking place in my brain. New neuronal circuits and pathways became more sensitive to firing again and established new and healthy memories diminishing Mr. Delgo's hold on me. Later on, during my professional and clinical experience, I learned that neuroscientists have amply proven that sensitization to a dummy like Mr. Delgo can effectively deal with pathological fear and neurotic behavior. The important lesson is that new brain connections take place when feelings, emotions, body and neuronal generated electricity are engaged when we commit ourselves to search the truth.

PSYCHODRAMA AND HEALING

I like to add that I attended several psychodrama sessions and it reinforced my conviction that to bring about permanent changes in our behavior, it must be an action guided therapeutic modality. We are an emotional organism and any effective healing process must address our emotions. During psychodrama sessions, we made use of role talking in which students, role-play introjects from the patient's past. Could be one or more introjects addressing the patients as if it were present in the room. There were also support members that participated in each session. It appealed to me because it emphasized engagement through active role-play and dramatic dialogue in a group setting. It is an existential healing practice. The whole body, mind, and if you please, the soul was engaged. I was consciously pedaling Mr. Delgo's dummy but unconscious conflicts were also brought up by role-playing introjects. My father, mother and dead brother were brought up during my sessions. Psychodrama is an experiential professional healing modality that provides honest and loving support. While I was attending New York University and after graduation, I set aside time to attend Puerto Rican Spiritism séance near my home residence. Quite often I became as excited in a séance as I was in psychodrama. Role-playing and talking therapy was practiced in both places. In both places, I felt unconditional support from every member attending sessions. With the help of the teachers' group and psychodrama, I learned to correct my own self-image. They helped me write down my own accomplishments which, I was blinded to by Delgo's memory. The group wrote it down as if it were a commendation and was signed by every group member. This was later done for every member of the group. At times we held sessions dedicated to supporting each other. We named things that we liked from each other which are often ignored by or not consciously recognized by ourselves. I recall a lady teacher whose drawing abilities were exceptionally good but, she felt it was natural for most people.

She was able to identify and incorporate that part of herself into her own. Cooking, writing poetry and short stories, as well as gardening, reinforced our group experience pedaling noxious memories. With group support, every one of us was discovering and expressing potentials we did not know we had. Pleasant memories supported by our group's approval became part of each one of us. Before I end up this chapter, I like to tell you that my adaptation strategy at the agency I was working for, paid me well. I read everything I could on Freud and upon the administration's recommendation, I was awarded a two-year scholarship. I chose New York University over Columbia or Fordham. My curiosity and clinical application of psychodrama and Spiritism séance to family and individual mental disorders led me to write a book. I followed patients attending Spiritism séance, their treatment and attendance at a local mental health clinic. It was revealing and very useful during my clinical practice.

RELIGION AND CHARLES DARWIN

During my sessions during Psychodrama, as expected, no mention of religion was made. However, at the beginning of every Spiritism séance, passages of the Bible were read. The words spirit and soul were used interchangeably. It was disquieting for me that the soul, as I understood then, was strictly a Greek invention or creature. On the other hand, the spirit was a Jewish/ Christian heritage, if you will. As I understood it, early Christians, especially apostles Paul and John, saw it fit to use words interchangeably to bring together Christian religion and Greek religious philosophy. I did not like it. The Bible I used for reference said, "God breathed into his nostrils his Spirit. I reasoned that in both Jewish and Christian theology, man's spirit comes directly from G-d. In opposition, in Socrates' religious philosophy, the soul comes from gods, angels or maybe another unknown source. I was going after the soul, the source of truth and wisdom. If you believe it should not have been confusing to me trying to integrate religion and philosophy into my personality; to understand Genesis according to this author. I would like to recommend to you to listen carefully to television evangelist. He writes, "God formed Adam from the dust and clay of the earth… breathed the breath of His own divine life into his nostrils, so that the man became a living soul."[20] He goes on to explain that G-d's breath is the spirit, and is thus G-d himself residing within the vessel of clay, the two together constituting a living soul.

So, I reasoned, there is Adam, the man of clay, G-d, his spirit and a soul. I said to myself, good-by monotheism. I felt that the best thing for me was to set aside the soul and all religious inquiry and concentrate in the here and now. I was miserably too small and ignorant to go into abstract and complicated subjects. To bring

---

[20] Alan W. Watts, Myth and Ritual in Christianity, Beacon Press, Boston, 1972, p. 50.

rest and peace to my mind, I took Charles Darwin, On the Origin of Species and underlined some sentences without giving it much thought about it. Here are some: "Never to forget that every single being around us may be said to be striving to the utmost to increase in numbers...the dependency of one organic being on another...is often the case with those which may strictly be said to struggle with each other for existence...the struggle almost invariably will be most severely between the individuals of the same species, for...the same food, and are expose to the same dangers. The structure of every organic beings is related, in the most essential yet often hidden manner, to all other organic beings. Man can produce and certainly has produced a great result by his methodical and unconscious means of selection...man selects for his own good." I was attempting to justify my departure from the soul. I thought I was parting with the soul, but it finds its way to you. I still go to Darwin and read over again things underlined many years ago. It reminds me that despite the diversity of animals and plants on planet earth, we have a common ancestor. A wonderful cell that was capable of dividing itself into many more equal cells with the potential of forming a complex organ like our brain.

READING FROM THE KABBALAH

Attempting to follow a straight line as possible during my mind trip from childhood to the present, I shall proceed with a sentence from the Kabbalah. "Descriptions exist of the mystical sensation of the subtle ether or aura...by which man is surrounded, of mystical visions of the primordial letters in the heavens... and of invisible books that could be read-only with the inward senses."[21] Around fifty years ago I had a very vivid dream- I called it a vision- with a book with its pages in the blank. During the dream, I was taken downstairs to a study room by an adult man whom I suspected was my double. When I protested that the book, larger than any regular size book, was in blank, nothing written in it, I was told to look deep into it, that everything I was looking for, all knowledge and wisdom on earth was there. In part, to justify my own curiosity and self-discovery, I have decided to look at it with my senses. I will use my eyesight, my hearing, smell, touch, and my emotional self in conjunction with my potential to understand the abstract. I will do the utmost to stay on target and learn the most I can about my body in flesh, emotions, and mind. I will do the best to leave Freud, Jung and Eastern Religious philosophy in its corresponding shelves. Regarding my dream, I like to quote from Karen Horney; she states, "Dreams...give voice to our strivings, our needs, and often represent attempts at solution of conflicts bothering us at the time. They are a play of emotional forces rather than a statement of facts. A dream is not understood until we can connect it with the actual provocation that stimulated it."[22] Doctor Horney wrote this book over fifty years ago and most students of neuroscience to date agree with it. Dreams are attempts by my brain to solve the problems of the day. It provides

---

[21] Kabbalah, Gershom Scholem, New American Library, New York, 1974, p.181.
[22] Karen Horney, Self Analysis, W.W Norton & Company, New York, London, 1942, p.163-4.

74

a solution to my quest for meaning in my life. Her recommendation that I must connect with it; and for me, now- connection means emotional connection; it must be a feeling connection. It cannot be an abstract exercise while sipping a drink. Most dreams take place while under rapid eye movement. The brain is in high gear and the brain chemistry-neurotransmitters- are crossing synapses and activating neurons of all sorts. We are a chemical machine or factory ignited by electro-genesis. We must connect with the causal factor or factors that provoke our dreams or as in my case, a vision as I decided to call it. The dream takes place in my brain chemistry and electrical component stimulated or provoked by external stimuli. Therefore, it is up to me to find the meaning of the dream by supplying the remaining incomplete parts of the dream. My religious, cultural, social, ethnic and situational contributing factors are unique and very personal to each person. Even identical twins have different dreams with a different meaning.

One of the leading researchers on the functioning of the human brain, Antonio Damasio states, "When feelings occur there is a significant engagement of the areas of the brain that receives signals from varied parts of the body. The brain areas (receiving signals) include the cingulate cortex; two of the somatosensory cortices, known as the insula and S11, hypothalamus; and several nuclei in the brain stem tegmentum.[23] The tegmentum is located in the upper posterior part of the pons and midbrain; the insula is located in the upper part of the temporal lobe. The cingulated cortex is located above the corpus callosum and the two somatosensory cortices are found as if portioning the cortex. The hypothalamus is located under the thalamus. Damasio has done superior brain research on emotions and feelings maintaining the unity of mind and body. His numerous publications on feelings and emotions have broadened my narrow outlook trying to understand mind-body dualism. Not that I am out of the tunnel but I hope that the dim light that I see far ahead of me is not another confusing dead end. The things that provoked my depression were

---

[23] Antonio Damasio, Looking for Spinoza, Harvest Book, Harcourt, Orlando, Florida, 2003, p.96

not what I learned or failed to learn from Delgo; it was the emotional experience that placed my whole body under siege by an abusive person. It is the emotion and the subjective state of feelings that break you down ending in neurotic behavior.

STEVEN ROSE, MY FRIEND FROM OVERSEAS

Through quotations, I will introduce to you an outstanding brain researcher who has made a very significant impact in my life. I had an earlier version of his work in which I lost. What impressed me the most when I first read his book was its simplicity and easy reading of subjects like the biochemical basis of human consciousness and personality. Even Santiago Cajal would be pleased with Steven Rose. I have underlined several times the following quotations in different colors because every time I read it; I found a deeper meaning to it. "It is the size of the brain relative to the human body. There are animals with bigger brains but our brain with at least 100 billion neurons and eight to nine times the number of glia cells that evolution, natural selection, and the environment made our brain the most complex and creative object in the known universe. The synapses and potential for learning can be counted in multiples of trillions. The potential that there could be more synapses in our brain than stars in the universe was awesome and unbelievable. He further wrote, "At the synapses between cells lies the choice point which converts the nervous system from a certain, predictable and dulls one into an uncertain, probabilistic and interesting system. Consciousness, learning, and intelligence are all synapse-dependent. It is not too strong to say that the evolution of humanity followed the evolution of the synapse.[24] Steven Rose's reliance on synapses for memories, learning, and intelligence was for me the opening door into the mysteries of the brain.

---

[24] Steven Rose, The Conscious Brain, Vintage Books, New York, 1976. P.79-81.

SYNAPSES, THE LOCI FOR CONSCIOUSNESS AND LEARNING

In a simple and short but profoundly significant fashion, this gifted researcher encapsulated what it is to be human. It is at the synapses that he placed an interesting system of consciousness, learning, and intelligence. The conscious man studied his own components parts in the process of learning; he incorporated the best from the outside and became the most intelligent organism known in the universe. Through the tools he has developed, he has gone deep into the universe in space and time. He has re-created the creation of the universe itself. This multicellular organism communication system, as Steven Rose puts it, "coordinates, control, receives sensory information and carries out appropriate responses. This set of properties demands a set of specialized cells.[25] This central intelligence receives, selects, analyzes, synchronizes and provides the most appropriate response at its disposal. These specialized functional cells in consonance with each other have developed an abstract yet unlocalized consciousness that sets self-direction and goals. This wonder-maravilla- has been developed and flourished in our planet earth. The basic components of our brain, its chemistry, and self-generated electricity, has a universal nature. Its beginning has to be found at the beginning of the universe, and its end is unknown. Steven Rose has taken me into an unimaginable trip with no end in sight. However, it is a trip that I have enjoyed every bit of it. Even my strangest dreams or visions, as I would like to call it despite my own doubts about visions, are beginning to have meaning to me. Insights from lay and professional persons have helped me gain a limited understanding of my personality. With the potential of trillions of synapses, my potential for knowledge seems unlimited. How sad it is to think that Mr. Delgo

---

[25] Steven Rose, op. cit. p.146.

failed his responsibility as a teacher and inflicted an emotional scar that took me many years to overcome.

On the specialized cells that Rosen referred to over forty years ago, neuroscientist, E.R. Kandel writes, "Every cell has in its nucleus…all the genes necessary to form the entire organism…a liver cell is a liver cell, and a brain cell is a brain cell because in each cell type only some of those genes are turned on, or expressed; all other genes are shut off or repressed…genes are switched on and off as needed to achieve optimal function of the cell."[26] This is the type of cell sophistication that is awesome to me. Our organism creates cells that become immensely specialized and develop a system to turn itself on and off as needed. It modifies its own proteins to achieve its goals. I cannot, but look in awe and admiration the wonders of my brain. It is the same brain that survived the assault on me by Mr. Delgo. My brain developed a system composed of my building blocks- proteins- to turn on and off genes as I needed. It is the self-regulating, self-imposed, unlocalized, but beautifully precise system of functioning that provided for my organism self-healing and remains a unique individual. Before I put Rose to rest and continue with either Kandel or Watson I would like you to share with me one more last quote. Please, join and support me in mind travel to Mr. Delgo's toxic memories in my brain. Steven Rose stated: "Memories are extremely difficult to erase once they are established…Memories in the brain are the most durable environmentally imposed individual characteristics…As the memory trace becomes fixed, (Mr. Delgo) a new process grows out from one neuron to another, making functional synaptic contact with it and therefore forming anatomically new pathway."[27] Here is the crux or vital point, the forming of an anatomically new pathways. The repetitive emotional abuse altered and modified my synaptic structure in a very negative way. My brain was growing up at a very critical phase of development but, in the

---

[26] Eric R. Kandel, In Search of Memory, W.W Norton & Co., New York, 2006, p.257.

[27] Steven Rose, op. cit. p. 235.

wrong direction. The synapses were transformed from sensitization to permanent anatomical changes.

It is very painful to remember but the memories that became permanent from Mr. Delgo's physical and verbal abuse to me during one of my most critical stages of development- ages 13 to 16- provoked anatomical changes in my brain. It prevented my brain from growing normally. Synaptic connections at the pre-frontal cortex, the last region of the brain to achieve maturity were prevented from developing normally and healthy. I left Junior High School with the impression that I was inferior and had below-average intelligence. Mr. Delgo had not wasted time letting me know that he was wasting his time on me. I was an object to be disposed of as he wishes without any consideration as a human being. During my eight grades, I asked the school Principal to send me to learn the carpentry trade. "I have to talk to Mr. Delgo and Mr. Ross" said the Principal to me. Mr. Ross was the teacher in charge of the carpentry program. I used to visit his home and befriend his son who was about my age. When I went back to the Principal he informed me that it could not be done. No further information was conveyed to me. Mr. Delgo left no doubt who was the boss in his classroom. I attempted to follow up upon the School Principal and Mr. Ross's decision on me. Two days later, I told my friend about my intention of joining his father's program. When I approached his father, he told me that the Principal had not talked to him. The following week I was sent to the agriculture program to work on the school farm. Mr. Delgo had pictured me a dirt laborer for life. However, this type of prejudiced mentality is not limited to Mr. Delgo. During 1964-65, I worked at an agency in New York City as an educational specialist, and to my surprise, I came across an unofficial policy among School Principals, teachers, and guidance counselors, that minority students were automatically placed in vocational school programs. They were discouraged from becoming anything else but janitors, dishwashers, street peddlers and low paid factory workers.

Obeying Authority
Without Questioning

I had learned to keep quiet and obey authority without question. My observations, whether correct or not, had to be sent to the recycle pail or kept in the silence vault. This tragic episode in my life was taking place during my most critical stage of bodily and mental development. I was surviving in a partially self-imposed cell as if I were incarcerated in myself made jail. However, my senses were at a high level of alertness and extremely cautious about what I said in public. Most communication was one way. Even my conversation with my brother and Mr. Clemens had to be restricted to listening; verbalizing or worse, challenging their opinion or interpretations of a particular subject matter, was a thought that had to be kept in closed doors and suppressed. The continued re-sounding voice of Mr. Delgo was inside my brain 24 hours a day, 7 days week, 365 days a year for at least three years, pounding over and over again inside my head was unbearably painful. It provoked abnormal synaptic excitability in brain areas like the amygdala, hippocampus and prefrontal cortex that are undeniable and lasting brain threats if not permanent damage. His rejection and anger projected towards me were always conveyed to me through his voice, eyes and minimum reference to me. I was his hated toy; he could discharge his frustration and anger without reprisal of any sort. Consequently, my social contacts outside my immediate family were conditioned by Mr. Delgo's memory. I had learned to hate Delgo the teacher, and to fear his accusing voice and menacing look. He despised and humiliated me; tortured and ridiculed me for his own pleasure. In retrospect, the hate that I felt against Delgo must have helped me maintain my sanity. My brain defensive tools must have prepared me to persevere and arm myself with powerful emotional and cognitive responses to withstand his daily verbal assaults. Self-talk like: keep on track, I am young, he is old, never surrender to the enemy and more, were part of my daily prayers. It was a duel between good and evil, right and wrong, justice and tyranny and, above all, the will to survive against all odds. While writing this essay makes me feel sad because this tyrannical and cruel person, like Hitler in Germany, was considered by many of his countrymen a semi-god. He was considered by intellectuals as well as laypeople the savior of the country. Adolph Hitler was projecting his wicked thoughts on defenseless people. Like Delgo, Hitler had to have a scapegoat to blame for his evil plans, meaning, burning human beings alive. In both cases, the majority of people around them approved their actions as if blindfolded. After so many years of cruelty, I rejoice that I am writing this essay. There is no doubt in my brain that I was born with exceptional adaptive survival tools to challenge and solve multiple problems of life.

BOOKS AND MORE BOOKS

My friends were books; books that did not put me down nor accused me of being poorly dressed, indolent and a mentally slow adolescent. I just could not do anything right, and if I could, I would not do it. I would have been accused of stealing someone else's work as Mr. Delgo often accused me of. It seems that I was on a self-destructive trip imposed by a toxic memory. However, after such a painful experience, my brain must have made intelligent decisions for my survival. I must add that despite my painful experience at school, I used to visit mama Tiny and friends from my early childhood. My friend that once was willing to jump out from a palm tree with me, was attending another school near his neighborhood. Occasionally, we used to get together and discuss a common hobby, raising rabbits at home. We became good at breeding the best of our stock. A nephew of mama Tiny had business connections with meat markets in the city. Our rabbits became a source of income for us. I bring this up because, despite Mr. Delgo's attempts to block my emotional and cognitive growth, I was always trying to protect myself from his destructive behavior to me. I was setting short and long-term achievable goals for me. The survival instinct was feeding my brain with tools to persevere and maintain a sense of emotional stability without dropping out of school. It was very painful for me to hear students-classmates- repeat the same words Mr. Delgo was using against me.

Now that I am convinced that Mr. Delgo's memories provoked anatomical changes in my brain, I hope you are also convinced how childhood traumas can provoke permanent brain anatomical changes. I spent many years of my life trying to reverse the damage done by toxic memories during my adolescent hood. When discriminated against during my trip to the United States based on race, ethnicity, language or cultural differences, I justified it based on my self-depreciation and poor self-esteem. It was a legacy from

Mr. Delgo. However, from books I had read and professional contacts, I had received support and valuable insight into my problems
and mental content. Besides, during my first two years working in
a newspaper cafeteria, I met honest and supportive individuals that
encouraged my efforts to help myself.

MAGGIE'S FAMILY SUPPORT

As I have said earlier, the landlady's daughter Maggie and her uncle never stopped pointing out my potential for growth and health. During my service in the U.S. Army, I enjoyed the friendship and support of my fellow soldiers. An exception was during my basic training. I was unfairly treated by the drill Sergeant. However, Mr. Delgo's humiliating memories popped up unexpected even during the best of my moods. I remember a Sunday afternoon when my landlady's daughter, Maggie, and I came out after enjoying a movie by a Greek comedian, Lou Costello. After the movie, we went to a recently opened cafeteria upon her invitation. It was the place to go after going to a movie and window shopping in the downtown Philadelphia at the time. Afterward, instead of taking the trolley back home, she suggested that we walk home. It was a cool, clear and dry evening. She was making all types of comments about the film. She tried to cheer me up and even held my hands when crossing the streets saying, she did not want me to get hurt by the passing trolleys. Thoughts came to my mind wondering whether she was in love with me. However, I proceeded to punish myself saying that she was a well-educated and intelligent young lady, well above my reach. It was Delgo`s voice inside me that belittled myself even in the best of circumstances. I was miles away from Mr. Delgo and the school where he punished me mercilessly. Space and time did not matter for toxic memories to continue to spoil my day. I was close to two thousand miles away from home and six years had passed since my last humiliating verbal assault in May 1946. Mr. Delgo's memory was there to remind me how unworthy I was of anyone's friendship, care and love. I was a mentally-slow country boy without any future ahead of me. The thought of Maggie holding my hands as a gesture of love or interest on me was dismissed from my brain as delusional or the product of a wicked mind.

There were other instances of support during my life experience that helped me overcome a childhood painful experience. I recall my last few months in College in Puerto Rico. I joined a group of students calling for the Independence of Puerto Rico. We had group meetings and it provided an opportunity for me to verbalize and direct my anger towards an object. Interesting enough, the group was directing their action against the existing established order in Puerto Rico. As I understood it, Mr. Delgo was part of that established order. Besides, I had excellent professors whom I could maintain off classroom intelligent conversation on political and historical issues of Puerto Rico. I considered them supportive agents of change that facilitated my emotional and intellectual growth. These are the sources, the integrating brain makes use of, to move us forward solving problems of daily life. On November 22, 1963, J.F. Kennedy was assassinated in Dallas, Texas. When I heard the news in my car, I went straight to my father's home and informed him of the tragedy. I said to my father that assassination was unacceptable in a democratic system of government. I added that the power of persuasion in public debates or, individual one on one conversation should be the only way to transfer or dismiss anyone from government. My father looked at me and then, moved towards the horse he was about to mount and said, "Obviously, there are violent individuals in the United States that do not care about the rule of law." I was going to tell him that there are violent people anywhere in the world but he went away on his horse. I was unaware but Maggie's brother, Paul, had taught me a lot. His arguments were valid and convincing. He did not lecture people; his method was an open discussion of any relevant issue. He was an excellent and sympathetic listener. I was trying to enter in an open conversation on violence with my father, but it did not work with him.

José in New York City, the Big Apple

For an unknown reason, I had resigned in my teaching position in a nearby town and decided to move to New York City in January 1964. I was academically equipped with two majors, one in education and another in history. Besides, I was registered in a Master degree in Public Administration at the University of Puerto Rico. In New York City, I found a job in counseling youngsters with emotional and behavioral problems at school and home. It included reaching out to their families and the community in general. I considered it the greatest challenge I ever had to test my body of knowledge in education and human behavior outside my community. It opened a great opportunity for me to help myself through the problems the youngsters and their families were going through. I would be able to share my feelings and thoughts with teachers, counselors and supervisors of the school system and our agency. Until that opportunity, my understanding of Freud, Jung and others, were self-taught. To my advantage, my supervisor was a professionally trained psychotherapist of the Freudian orientation. I made it my responsibility to learn Freud theory in its minimum details. I bought Freud's latest and most reliable translation and discussed it with therapist trained in Freud's theories. I began to explain everything that fell in my hands in Freudian terms. I made a very good impression on my supervisor and the agency administration. However, in reality, when I visited the families and their children at home, Freud theories became somewhat short. I could not explain with the youngster's behavior strictly on psychoanalytic theory. Something was missing and I began looking around to see if I could find the missing pieces to add to Freud theories. One of the books that attracted me at the time was Carl Rogers', On Becoming a Person. I remember underlining the following, "I found that it does not help, in the long run, to act as though I were not something that I am not." I thought it applied to me for several reasons but mainly because what I observed during my contact with families, both at home and in my office, presented a different picture, if not contradicting my understanding of Freud and his approach to human behavior in general. Another underlined sentence is, "The more I open to the realities in me and in the other person, the less do I find myself wishing to rush in to fix things." I understood from the above underlined teachings of Rogers that I was boxed in or caged in on Freud theory and I did not see my real self. I used to question myself saying, "How could I see and understand myself if I do not understand what the real self is? I was struggling with the soul, mind, body, self, death and introjected objects besides Mr. Delgo. Interesting enough for me now is that even molecular biologist and biochemists are tackling about the same subjects hoping to find an acceptable and scientific supported answer for many of those same problems of mine. Needless to say, but I was trying to take in more that I could possibly chew. I began to realize that there was so much to learn about myself and others that my thirst for learning would keep me busy for life. I felt at ease talking about Freud and Jung, but Behaviorism was catching up. Fritz Perls, Carl Rogers, Albert Ellis and others were also making inroad into the field of mental health. This was a time of social unrest, Vietnam War and civil rights protest and marches to Washington. The Black Panthers, The Young Lords, The K.K.K. as well as leftist, liberal groups and individuals that were protesting for the sake of protest were adding to my already confused reality. I was searching for the truth but my truth and that of most American at the time was wrapped up in layers of myth and unidentified sets of values.

MAGGIE IN PHILADELPHIA

My friends from Puerto Rico had come for summer training at a New York City University. They wanted to visit Philadelphia, the cradle of American Independence, as they put it. We went by public Greyhound buses, we left early morning on a Saturday and by 9 a.m. we were at Independence Park in that historical city. The first thing we went to see was the Liberty Bell; second was Independence Hall. I took a stride to see the old tree where Maggie's uncle, Paul, used to take me to meet his friends talking about world politics. To my great surprise, there was my friend Paul, now looking old with a long white beard. We ran towards each other and hugged each other as if we were family. When he asked me how I was doing, something got stuck in my throat. I could not answer him. He pulled me against him and said, "My family always loved you." Soon thereafter, I introduced the teachers to him and explained him the purpose of our visit to the city he loves so much, and I have never been able to take off my mind, Philadelphia. Paul addressed them in Spanish saying, "mis amigos camaradas." The chemistry between them seemed to work very well. Paul, pointing to a corner building very close by, said to me, "Maggie has a part time job out there, go and say hello to her. I will take care of your friends." I guess I did not want to believe him so I did not move; he pushed me lightly and repeated, go ahead, it is okay. I walked inside the building and there was Maggie. She was moving a framed picture from one place to another when she looked up and saw me. I became anxious and hesitated to take a step forward to meet her. She looked around before she found a place to lean the frame against. We walked towards each other without saying a word and remained hugging each other for a little while. My mind was overwhelmed by her memories. My throat was stuck with something I could neither swallow nor spit out. With a small handkerchief she wiped away a tear or two. After a short while in silence, we talked a bit about ourselves. We touched on evenings we had spent together.

I told her I was married and live in New York City. She told me she was divorced and has a son. When I told her I had to leave and join my friends, I noticed her eyes watering and she said to me, "You always found me too old for you, didn't you, Jose?" Of course, I could not answer her, but I hugged and kissed her like no one before. I whispered into her ears, "I always loved you." I walked out and never saw her again.

Mind-traveling has provided me very rich, happy and enlightening experiences that have supported me to overcome painful experiences. When I left Puerto Rico before I had reached age 19 and reached Philadelphia believing I had landed in New York. Maggie, her mother and uncle, Paul, more than compensated for Mr. Delco's sour behavior. They came into my life during a critical period. Consequently, my brain connections, especially, in the prefrontal cortex and its projections with and into the limbic system must have provoked healthy changes in my personality. After serving two years in the Army, I was able to get a G.E.D. High School Diploma and I was admitted to College. After my graduation from the University of Puerto Rico, my plans were to return to New York City or Philadelphia. Interestingly, I did not go back to live in Philadelphia until 1977 to 1980. I did not look nor did I see Maggie or her uncle, Paul. I repeat, just a couple of years in Philadelphia with loving and caring people made me come for more. Most of my brain changes took place while in Philadelphia during my relationship with Maggie and Paul. They were listening to my feelings silently, and I was responding to their love and care in a similar fashion. It seems that there were no demands from either of us. I have traveled to Europe, Mexico, Puerto Rico and mostly in the United States, but my mind always returns to Paul and his family.

PIAGET, AN ALTERNATIVE TO FREUD

Before I proceed with the phase of my growth and development, I would like you to know that besides Behaviorism and psychoanalytic theories, Rollo May's Existential Psychology and Carl Rogers; I had considered Piaget an excellent alternative to help me resolve my problems and apply it to families and adolescents I was working with. However, reality around me, meaning people I worked and socialized with were of the Freudian circle. Above all, I wanted to survive, and accommodation always pops up in my mind. Nevertheless, psychodrama and less known therapy group sessions were held in the City and provided an opportunity for me to learn from it. Also, I spent two years under therapy with a non-Freudian European trained psychologist. He called himself an eclectic oriented therapist. With my own clients, I was teaching them through role-modeling and other appropriate techniques to get in touch with his/ her body's feelings. Painful feelings of sadness, loneliness, anxiety, rejection, chest tightness, as well as joy, accomplishments and sharing are felt and identified as existing in the body. Through awareness, the client learns to unlock and release energy available to the conscious self for continued growth and happiness. Good health is dependent upon normal flow of this energy. The lessons and insight gained from outside contacts; I was introducing it into my practice besides my Freudian orientation. I was always looking within my experiences, the root of my own conflicts. I became aware that if I deny my own discomfort or the secondary gains I might have received from my own maladaptive behavior, I was impoverishing myself. I had taken upon myself to fill up the empty book I was shown during a dream I called a vision. It was not a diary, but learning experiences put into practice during group sessions, private therapy sessions and experience with my clients. It does not have to be said that my insights and learning experiences were validated against professional and academic make up. I used to underline my own empty book with preferred quotes like the following one, "Growth demands awareness and an emotional commitment to make a constructive rational decision from among multiple alternatives. Our knowledge of freedom is the knowledge of our own necessity to choose from among various alternatives in moment by moment sequences."[28] Another teaching lesson from F. Perls that never left my brief case was the following, "Awareness per se, of and by itself, can be curative. Because with full awareness you can become aware of the organism's self-regulation. You can let the organism take over without interfering, without interrupting; we can rely on the wisdom of the organism"[29]. Let me add that Perls' book provoked my curiosity to attend a free public demonstration and later on, I spent four years in a Gestalt School and Teaching Institute mastering its teaching and helping myself.

[28] Wilson, J.A., Mildred C. Robeck, William B. Michael, in Psychological Foundation of Learning and Teaching., New York, McGraw Hill, 1974, p.188.
[29] Perls, Fritz, Gestalt Therapy Verbatim, Utah: Real People Press, 969, p.16.

P.A.R.T.M.

The origin of pedaling (kicking) and removal of toxic memories (P.R.T.M.) that I practiced and our group of teachers has a very interesting history. When I was around eight years old my friend Chito, of about my age, used to draw the face of my father and his father on the sand when they punished us. I do not remember whose initiative it was, but we had a good time stamping their faces with our bare feet. We made believe that we pulled their nose and ears just as they pulled ours. When our emotions became excessively high enjoying our revenge, if you please, we took a shrub and lashed out their faces until it disappears from the sands. Occasionally, we used to steal oranges from a neighbor's garden. The neighbor used to complain to my father and Chito`s father culminating in our punishment. We cleaned a space in the sand and drew a whole man's body with both of our father's name. We cut a branch from a nearby tree and enjoyed the best revenge ever. The neighbor was a very old man we used to make fun of. Chito`s memories were always popping up in my brain. Once I taught third and fourth grade in a rural school. It was half a day for each grade. By coincidence, I watched how boys used to draw faces of men and women and then, make holes on it with pencils. I associated their behavior with mine when I was eight years old. You can observe small children doing the same with toys. It seems that aggression has its roots deep in our genes.

During my Gestalt therapy sessions, whether in group or individual, what helped me most was the freedom I gained through the support from the therapist and the group to unlock and release those held back energy. I used the newly discovered energy to raise the level of emotional touch to get my whole organism involved in the destruction of my past introjected toxic memories. The higher the intensity of my emotion, the higher the excitation level and sensitivity in each neuronal cell and its corresponding synapse. Now I can add that neurotransmitters like dopamine and glutamate, among others were abundant in my brain during this level of excitation. The synapse becomes sensitize during this level of emotional arousal. I suspect it would be equivalent to neuroscientist, Eric Kandel introducing serotonin to promote dendrite growth. It would provoke the birth or sprouting of new dendrites or spines at axons terminals. I suppose the bigger the excitement, the larger the number of dendrites and spines. It is translated as building new neuronal bridges.

ESPIRTISMO, GESTALT AND PSYCHODRAMA

Looking back to my experience with Espiritismo séance, there were mediums that recommended to the introjection. The victimized individual, to take his pant's belt off or any other available and useful tool, place the imagined bad spirit on the table and whip it. The medium claimed that there were spirits that needed to be disciplined. He further explained that spirits had to be taught good rules of behavior. It seems to be in consonant with Allan Kardec; his doctrine teaches that some spirits do not know more than living individuals. Like in here, is in heavens. During psychodrama sessions, there were therapist that called upon the whole group to verbally admonish and reprimand the introject behavior. As in Espiritismo, group support played a leading and decisive role restoring human health and growth.

Learning experiences from psychodrama, P.R.T.M. therapy and Espiritismo led me to appreciate the teaching and therapy of Gestalt psychology and treatment modality. Its founder, Fritz Perls was a Jewish psychiatrist that left his country in time before Hitler and his criminal gang killed five million of his people. Perls writes, "In Gestalt therapy we therefore require that the patient psycho-dramatically talk…to the introject. He (the patient) has to transform his thought about the past into actions in the present which, he experiences as if the now were the then. He cannot do it merely by re-counting the scene, he must relive it. It is insufficiently merely to recall a past incident, one has to psycho-dramatically return to it…We ask them to localize the pain and to stay, or sit, or lie with the tension. The concentration technique provides us with a tool for therapy in depth … by concentrating on each symptom; each area of awareness, the patient learns several things about himself and his neurosis."[30] Now you can imagine my impression when I read his book for the first? I felt not only at home, but as I understood it, my past experiences were partially validated by an internationally known treatment modality. From childhood, I was psycho-dramatically re-living an experience in the present time. I was talking not only to Mr. Delgo and my father in the present time; I was addressing the bad object I had introjected. What destroys you is not the flesh and blood father, but the bad image you introjected and provoked changes in your synapses and neuronal structural changes. Most therapists understand how hard it is to deal with post-traumatic stress disorders. I was unaware that Cognitive Behavior Therapy was also another therapeutic approach most applicable to neurosis. Moreover, Gestalt offered me something that was unique and useful in my quest for self-identity and growth. Among my colleagues and friends, it was more appealing than C.B.T. At the time, I was working in a hospital with an outpatient psychiatric clinic and I believed the whole staff was of the Freudian orthodox orientation. However, the chief psychiatrist, a Freudian himself, used to encourage me by saying, "I met Dr. Perls once during my visit to Cuernavaca in Mexico; he is a very interesting person." I took his comment as supporting, if not, not rejecting my enthusiasm with Gestalt.

---

[30] Fritz Perls, The Gestalt Approach & Eye Witness to Therapy, Bantam Books, New York, 1973, p. 67-69.

GESTALT, THE GOLD STANDARD THERAPY

From now on, Perls and his psychotherapy modality were the gold standard of my thinking and behavior. I used to recite some learning and teaching lessons that I had to master like: Therapy consist in rectifying false identifications. In therapy, we have to re-establish the neurotic's capacity to discriminate. We have to help him to rediscover what himself is and what is not himself; what fulfills him and thwarts him. We have to guide him towards integration." [31] In all honesty, during my practice, my guiding principle was Perls therapeutic model, but modified to suit my needs as well as my client's needs. During my therapy sessions with my clients, I integrated learning lessons and tools that fit best every person I worked with. One of the attractions of Gestalt that called my attention instantly was turning the past into the present. The here and now that Fritz demanded from his clients while in the hot chair. You were motivated to go into your feelings as soon you took the seat in front of the group. You had to feel the pain inside you and not just a fantasy of the object hurting you. It was the self-contact that I had to be aware at all times. Gestalt refers to the I and you, in the present time, not a projected abstract like the pronounce it. When Mr. Delgo was punishing me in front of the class I would say, "I am dissolving into nothing and disappearing from everybody's sight. My legs and body is disappearing into nothing. I am invisible." I moved from the feeling to the emotion itself. I went from the introjected object, a fantasy, to the emotional state of self-destruction, if you please. Consequently, I could rebuild myself through the integration process of creating reinforced synapses or creating new dendrites and spines. I want you to visualize how I felt when I had to appear in front of people. When I had to

---

[31] Perls, Ibid, p. 42-43.

appear in public, my first defense, although pathological, was that of not existing. The emotional load I was carrying was unbearable, but nevertheless, I was benefiting from it. It was through Gestalt that I became aware of myself and the game I was playing on me. During the process of self-discovery, I learned self-integration. Learning, as I understood Perls' was not memorizing a set of rules of behavior or shoveling the unconscious into consciousness while in the comfort of a couch. It is an action process whereby your entire body is engaged. I discovered that what I was doing to myself was what I wanted to do to others, particularly to Mr. Delgo. My goal was not only to get rid of Mr. Delgo from my brain, but build myself new brain connections. Of course, at the time Perls was writing and practicing his Gestalt psychology. The concept of the birth of new dendrites and spines existing in our brains was science fiction, if at all. However, the Canadian psychologist, Donald Hebb, classic conditioning of Pavlov and Skinner were preaching something different and new. I understood that growth is not a linear process, but a bouquet of experiences with freedom to build your own garden, and the garden is yourself. Leaving your garden and coming to your garden when necessary and appropriate is learning.

Through Gestalt therapy I had my empty book from my dream almost filled up. My experience as a therapist was in an outpatient psychiatric clinic. These ambulatory patients were taking care of themselves fairly well. What was waiting for me during my next phase of learning and development impacted me like a thunderbolt. I began to work in an emergency admitting psychiatric unit. It provided inpatient and outpatient psychiatric services. Ambulances were bringing in all types of psychotic patients, some with symptoms I had not seen before. The professional and clinical intervention in this type of setting was almost totally different from my previous experience. Patients coming to the emergency room needed immediate nursing and medical attention. A lot of them were violent patient that needed a large dose of antipsychotic medicine like Thorazine, later, improved by Haldol. The psychiatrist, nurse and security officer were in the forefront of every new arrival at the hospital. My intervention was limited to helping with family members and their needs. It was no small matter. In many cases, working with family problems were a bigger challenge than the heavily sedated patient now amenable for an interview. Some family members were very angry at the patient, they had brought in because they had been victims themselves of the psychotic behavior. Family members wanted us to take the patient out of their house and place him anywhere, anything but sending him/her home. A big disadvantage for me was the scarcity of community resources, especially for mentally ill persons. Most patients brought in by ambulance, which was the vast majority, were from the very poor neighborhoods. They had family members like grandparents and grandchildren at home complicating the situation even more. For us, non-medical staff, the psychotic patient's problem was transferred from the emergency room to his home and family.

The emergency room is not a place for follow up, but as it says, it is an emergency intervention. I do not know if you can imagine how frustrating it can be for anyone working there. The demands are overwhelming from the inside as well as from the outside. You want to help, but you must move on. Poverty, discriminating practice, corrupt officials, shortage of housing, an outdated educational system, violence and drugs; just to mention what comes to mind now were out of my comprehension. The staff at the emergency room, with few exceptions, was in a revolving door. They used to ask me, "How can you take this punishment?" However, for me, it really opened my eyes and mind, excuse me, I should say, my brain. I had read books in human behavior and mental illness, but as Perls said, "Do not fantasize, live the experience and learning takes place."

GILLES DE LA TOURETTE

I will tell you one of my most painful experiences in that emergency room. One Saturday afternoon, around 3 p.m., an ambulance came in with a mildly agitated patient; it was a normal thing for us. I was interviewing a lady from Puerto Rico, jointly (,) in Spanish, with the chief psychiatrist. The newly arrived patient was forcefully helped to sit down and tied to a heavy chair we used to have in the emergency room. While I was going to go to my office, in just a flash of a second, I saw his face. It was the face of a medical doctor I used to refer clients to, from my previous job. Occasionally, he used to invite me to his office. He lived alone, and was divorced; had two children living in California. I rushed to him and observed, he was moving his head rather violently right and left, making strange noises and to my great surprise, the polite and well behave friend was using foul language. That was not the professional person I used to talk with. I went to the chief psychiatrist, who was ready to see another patient and told him my bewildering surprise tied up in a wheel chair. He asked an intern to take care of the patient he was going to see and said to me laughingly, "Come, you and I will see your friend, I know his problem. You will learn; it is a very rare case you ever see in a public emergency room. I wheeled my friend to my office, the psychiatrist saw him briefly, went inside the medicine cabinet and injected him a sedative. I told the psychiatrist, how I came to know our guest doctor and he said to me, when you can; look at the D.S.M. IV under Gilles de la Tourette. Then he added, "He trusted you; the staff in this emergency room trusts you, too." "Thank you," I said to him. After an hour and half or so in observation, we went together to see our guest again. He seems to be my old friend again, polite and well composed, no foul language, no shaking of his head; he appeared to be a normal person again. An injection, a chemical had gone to his brain through his blood, fantastic therapy. The chief psychiatrist said to me, "It is all yours, Jose." Going against all protocol, I took my friend to the hospital cafeteria; he explained to me how he had ended up in our emergency room. He was very grateful for our services. Actually, I took him to the cafeteria to satisfy my professional curiosity and doubts. I wanted to make sure to my satisfaction that it was alright for me to send him home.

An intra muscular or intra-venous injection had brought back my friend to his normal self. This miraculous cure had taken place many times before, but with people I cannot relate to. I was not dealing with the memories of Mr. Delgo, Freud, Jung or Perls. The chemical injected whether orally, muscular or intravenous was something I could see or touch if I wanted to. It went to the brain through the blood circulatory system, something that could be seen and touched in the laboratory. My next challenge was to try to understand the brain anatomy and physiology. I had read an introductory small book on Santiago R. y Cajal's neuron theory; it was very interesting and a very complex subject for me, a social science student. Along with Cajal, I read over and over again, Steven Rose's The conscious Brain, The Neuronal Man by Jean-Pierre Changeaux, Gordon M. Shepherd, Richard F. Thompson, A.R. Luria, Hynd and Obrzut, Scientific America and the New York Times science section replaced all other readings. About same time, The Broken Brain by Nancy C. Andreasen definitely turned my head upside down. Luckily, I had a friend at the emergency room whose Biological Psychiatry helped me a lot. While Perl's was my guide while practicing psychotherapy, neuroscience has become my enlightening hobby and best friend.

Parkinson and Alzheimer Patients

Within a span of three months after my experience with the medical doctor in the emergency room, I was assigned in a 47-year-old Caucasian divorced male. He had a double diagnosis of a mild Parkinson's Disease, and Depression. I had a part time job in a nearby outpatient clinic. For a reason I never questioned, the administrator said to me that it was her opinion that I was well qualified to work with that patient. I was going to burst into loud laughter, but the way she sounded did not call for that type of behavior on my part. To the best of my knowledge at the time, a Parkinson's patient belonged in a neurologist office. It was easier for me to talk with the medical director and I went to him with my set of complains. He said, "You are partially correct, but that patient needs much more than just medication. I concurred with the administrator" Well, that was it, no room for more arguments. It was kind of love and hate situation over my part time job. In about two and a half months, I was assigned to a Hispanic 53-year-old, separated lady with a dual diagnosis of Alzheimer's, and Depression. I was already working with a paranoid schizophrenic and a bipolar patient. If you do not begin to think that someone is picking on you, you deserve a gold medal. The Alzheimer's patient was the last straw that broke the camel's back; I had it. Assigning me to four chronic patients in a very short period of time was like forcing me out nicely. At least, it was the way I understood it. The medical director came from northern India and our chemistry seemed to mix fairly well. I rang his phone and asked him if he had a few minutes to spare with me. He must have read and heard very well my anxious voice and said, "Yes, come in right now."

I knocked on his office door and he opened it for me. I became somewhat apprehensive; he had never left his chair to open the door before. Things became even more tensed when I saw him prepared a cup of coffee for me. What is he up to now? He must have known about my new assignment, a patient with Alzheimer's. My brain was busy forming multiple designs to solve or calm down my anxiety.

Silence reigned for a short while. I took a few sips of the delicious coffee while he looked at me, half smiling and said, "You have another very interesting patient being assigned to you." I noticed my heart bouncing against my chest, my brain must have been receiving tons of blood, glutamate and other excitatory chemicals forcing neurons to fire like a fourth of July in downtown Manhattan. I was going to say he was right, they were interesting patients, but I had enough of it. I was very much interested in exploring the biological basis of mental-brain problems, but I needed to practice Perl's psychology. There were other clinics where my services would be welcome. Back on his chair, he placed his right hand on a stack of charts, meaning patient's records, and said to me, "I handle all types of psychiatric problems in this clinic. It helps you work with psychiatric and neurological problems, too." I did not respond to him. I was trying to translate his message to me before I react to anything else. He observed my emotional state. He was handling psychiatric and neurological problems, but that was his training. I was going to tell him not to throw that b.s. on me; but I reconsidered and kept my mouth shut. He was not a person to do that. I did not know how to respond to him. He got up from his chair, walked in front of me, checked the door to see if it was locked and said to me, "Three days ago, we had a meeting at the central office. The dentist next to us is leaving and we will take over his space. We will bring in more staff and you will be their supervisor." I did not say anything; he approached me, shook my hand and said, congratulations! When I was ready to leave his office, he added, "Let's keep it a secret between you and me." Things were moving faster than I had planned and in a direction I had not anticipated. The following week the administrator called me into her office as soon as I got into my office. I must have wanted to make a good impression on her because I took two books with me. Both were related to Parkinson's and Alzheimer's disease, I placed it on top of her desk. She looked at it and with a broad smile on her face said, "I knew it, I knew it, there is no problem you cannot solve." I joined her laughing saying, Alzheimer, Parkinson, schizophrenia and bipo-lar, it sounds like the plagues of the Apocalypses. Still laughing, she looked at me and added, "Jose, you can do it; the dentist is moving out and we will be moving in. You will be in charge of supervising the new staff in that section. You have my full support." She asked me to take my books with me and leave her office because she had an urgent meeting and had to go running. Was it a blessing or a curse, I was already a supervisor at the hospital I worked for. Now, without the asking, I was saddled with more of the same.

A Blessing?

I took it as a blessing with reservations. It motivated me to continue looking deeper into my brain. I needed to study and learn more about the physiology, anatomy and chemistry of the brain. I needed to understand why dopaminergic brain cells in the substantia nigra stopped producing and releasing dopamine as in Parkinson's disease. I was caught in the seemingly paradoxical issue of dopamine in Parkinson's and schizophrenia. In the former, it is a deficiency of it while the latter, it was blamed for an excess of it. An excess of dopamine receptor D2 in the striatum, -caudate and putamen- seem to attract or provoke the release of more dopamine from dopaminergic neurons, basically in the brain-stem. New research has elaborated and clarified the issue. Now sadly, I reflect how orthodox psychoanalytic literature authored by world known practitioner used to blame the mother, the bad mother as it used to be called, for neglecting her child and consequently, creating schizophrenia. Dopamine was also described as the addiction neurotransmitter along with endorphins. It seems we need someone to blame for our own shortcomings. Alzheimer's disease was the biggest challenge I had ever encounter. These four plagues of the Apocalypse are brain diseases and not mental disorders that I could apply Gestalt psychology or other therapeutic modality I had already practiced. Besides, brain diseases and mental disorders, what I really wanted to get into is memories, especially how can I destroy toxic memories. I used to tell myself, "You are on the right track, Jose; you are learning about the brain, your brain, keep on target." To keep up with the latest discovery, I bought the most recent research books on Biology and Biochemistry. I was lucky enough to get a copy of The Gene Concept by Natalie Barish translated into Spanish in 1968. It brought me back with the work of G. Mendel. However, what was eye opening and neurons firing for me were the D.N.A. and R.N.A. molecules. Further in line was James D. Watson and Francis Crick double helix structure of D.N.A. chain.

THE CORTEX IS NOT JUST
A COLLECTION OF...

In the meantime, I used to go back to the Steven Rose's book trying to fill up my empty book and I underlined concepts I found interesting and relating to me. "...The cortex is not just a collection of cells making chance connections with their neighbors, but rather there is a regular and precise wiring diagram involved, by which the connections between individual cells are specified."[32] Later on, I would learn that the precise wiring and specification of cells responded to signals proceeding from the nucleus of each cell. There is a genetic code specifying each cell in your body. There is no confusion as to what is going to be a liver cell or a brain cell. However, despite this precision, not even identical twins are exactly identical. Epigenetic plays a role in determining who you really are. Naively looking for the self-healing brain, I underlined Rose's teaching on plasticity. "If the left hemisphere has been dominant and is damaged, then almost complete takeover of the function can be made by the equivalent region in the right hemisphere." I have the impression that much of today's physical therapy, meaning rehabilitation, is based on this principle. One more quote from Steven Rose I hope not to offend anybody. In reference to memory storage he wrote, "The doctrine of memory storage occurred...was advanced by Donald Hebb...If synapses are modified or activated during learning the implication must be that the actual anatomy and biochemistry of a brain in which, a particular item has been stored must be different in some way from that of a brain prior to such storage." [33] Hebb synapse has been amply documented in laboratory by researchers worldwide. I hope I will not be accused of exaggerating Mr. Delgo's three years of torture.

---

[32] Steve Rose, op. cit. p.70
[33] Steven Rose, op, cit, p.237

Three years of most daily torture not only sensitizes a synapse, but provokes spinal growth modifying brain structure. Adolescents need guidance and love to overcome the turmoil of emotions and brain wiring to become healthy adults. We are extremely vulnerable during this period of personal development. I must add that I had excellent teachers who became my friends for life. Delgo was a very painful thorn in my head, but the will to survive and ultimately stop his grip on me, is a joyful memory.

Before I proceed with recent brain research I would like to quote learning tools, concepts that I have made use of many times. "The principal features of the anatomical and functional organization of the nervous system are preserved from one generation to another and are subject to the determinism of a set of genes that make up what I have called the genetic envelope. This envelope controls the division, migration, and differentiation of nerve cells; the behavior of the growth cone; mutual recognition by cell categories; the formation of widespread connections, the onset of spontaneous activity..."[34] *Changeau* is referring to the genetic envelope that resides in the nucleus of the cell. This envelope, the genome, was the subject of a great national challenge between Francis Collins and Craig Venter. Hopefully, the sequence of the human genome will help us discover genes responsible for many of our brain diseases. Ever since R. Cajal`s monumental work on the brain neurons over a century ago, the interest and enthusiasm by brilliant scientist to make the unconscious brain accessible to all of us is to be a reality. Cajal had established that the neuron was the basic structural and functional unit of our thinking organ. His observations led him to postulate that the neurons axon communicate with another neuron through very specialized endings later known to us, as synapses. Actually, the axon buttons or endings discharge its contents, namely, the neurotransmitter in the synapse. The synapse is the space created between axon's endings and dendrites of another neuron or its body. One neuron may make eight or ten thousands connections or synapses. Communications between

---

[34] Jean –Pierre Changeau, Neuronal Man. Pantheon Books, New York, 1985, p.227.

neurons flows in one direction from pre to post-synaptic neuron. Cajal's insight into the neuron functions gave way to modern molecular biology, mind biology as well as memory formation science.

J.B. WATSON AND F. CRICK

In 1953, James B. Watson and Francis Crick announced they had discovered the door that led to link between the living and non-living world. They did not tire by giving us the double helix, but followed on molecular biology to find, along with other scientists, and the genetic code. How amino-acids are formed from codons or triplets of nucleic acids to form the basic building blocks of our organism, and proteins. Within Cajal's discovery of the neuron as the basic unit of communication of our brain, scientists began to look for the cause of brain diseases between neurons signals and structure. Of course, my attention was focused on scientists' discovery of medicine that referred or related to patients I was following in the clinics I was working for. My greatest concern was Parkinson's, Alzheimer's, schizophrenia and bipolar diseases. These were the cases I felt more vulnerable. I felt not only incompetent to deal with it, but to this date, despite national and international research, much remains to really fully understand and find a cure for these brains' maladies. Besides our genes, there is the environment that definitely contributes to our illness. The message is that scientists alone cannot do it; it must be a concerted effort for all of us. We share this planet together.

My clients demanded all my attention and energy. I was giving them the best of my professional and clinical experience. It required that I study in most details every research paper, magazine and book I could hold in my hands. It was not a five days a week job, Saturdays and Sundays were fruitfully used for seminars, conferences and in vivo practice with outstanding specialists in the field. You may wonder what happened and where did I place Mr. Delgo's memory. You must remember my learning tools from Rose's and Perls' both alerted me that those memories were extremely difficult to erase. I never gave up, I was always learning. Later on, I learned from Eric Kandel that… the number of synapses in the brain is not fixed, it changes with learning…long term memory (Mr. Delgo) requires anatomical changes. Repeated sensitization training causes neurons to grow new terminals…whereas habituation causes neurons to retract existing terminals.[35] I am attributing challenging and pedaling Mr. Delgo during College, together with my friend teachers, group therapy and Perls' psychology helping me grow and change synapses. Growing new synapses through the growth of new dendrite spines took me to a new level of self-understanding and cognitive development. New, useful and supportive memories replaced old ones.

[35] Eric Kandel, op, cit, p. 215.

MY RECOLLECTION OF THE DEFINITION FOR MEMORY

Memory is a mental function for recalling past experience. For the sake of curiosity alone, I checked Oxford Dictionary, 2005. It says, "Memory is the faculty by which, the mind stores and remembers information." I went for the meaning of faculty in the same dictionary and found that it is a basic mental or physical power. Power as described by the same source is used as in "the faculty of sight." This is Oxford Dictionary of 2005 and not a XVII century dictionary if it ever existed. Not a single reference to the brain was made. According to Oxford, it is the mind that stores information. In reference to faculty, it seems they don't know where to put it, mental or physical power. I believe Benito Spinoza followed later, by Voltaire were strong enough to give the brain its legitimate place in memory formation. If they did not fully say so, the insinuation was there. Throwing the soul out of the brain in public might mean ex-communication, burning in the stake or the guillotine. In antiquity, it was the soul or gods that gave or took away from you, the faculty to remember memories

Recently, memories have been classified and described among others, into working memory, which is basically immediate memory. Suppose you are having a conversation with your friend about a conference you attended yesterday on learned fear. You ask him to call Pete at 000-000-0000, his telephone number, which consists of seven numbers. Pete will fax the entire conference to you as soon as you contact him by telephone, your friend added. You pull out your cell phone and call Pete. You retained the telephone number for that short period of time. Lo and behold, that was the memory working for you. That working memory for seven digits might be forgotten if you do not need to use it again. Before I jump into the memory or memories that have been bugging my mind, or should I say my brain for a long time, are long term memories. Note of relief, I am not to mention more noxious and painful memory of mine with the teacher Delgo because I may become a noxious memory to you. Besides, I have had other caring and loving teachers that guided me well. Some researchers add explicit memory to the list of memories; explicit is declarative memory. When we are recalling a past event, it is dubbed, episodic memory. Besides, there are implicit memories that refer to automatic-like behavior as when you are riding a bicycle. When I first tried to ride a bicycle, there were no tricycles to get the first lessons. I learned by falling to the ground several times as well as bumping into all kind of obstacles that happen to be in my way. Working on my memories with my analyst while in training, I discovered my anger and fear towards my mother. During early childhood, I had heard from my oldest sister and my mother that my elder brother had died when I must have been between 18 to 24 months old. My mother transferred me to my eldest sister. She had to quit school to take care of me. My mother claimed that my brother had died of chronic anemia and asthma. Most likely, he died of the disease known as thalassemia. Three of my siblings and I have been diagnosed of having the thalassemia traits. A nephew of mine died as a child in New York City of the disease and two more girls from another sister are receiving treatment in Puerto Rico. I will get back to these memories as soon as I can in this essay.

JOSEPH GALL

Continuing with memory, I recall once reading a book by F. Joseph Gall. The name stuck by my brain because there was a fellow student who used to call me Joseph Gollo. I understood Joseph Gall, as saying that all mental processes, meaning that all that was taking place in my head were biological. As I understood it, biology and brain were one and the same thing, meaning that he left out the soul. Furthermore, Gall postulated that the brain- cerebral cortex- has separate and distinct compartments or regions for specific brain functions. For reasons I still do not remember, I liked his theory. Further-more, when his theory contradicted R. Descartes dualism, a theory that had been pasted in my brain with crazy glue, I felt some relief that I was not the only one struggling with cogito ergo sum. Gall was ahead of his time for about two hundred years although, his theory received significant modifications. Gall had assigned different mental functions to specific regions of the brain. The brain functions were based on bumps on the head. These speculative observations were subjected to doubts by science oriented professionals in the field. Besides, there are brain regions like the amygdale, hippocampus, limbic system, prefrontal cortex and hypothalamus with specific functions. Later on, two physicians, P. Paul Broca from France and Carl Wernicke from Germany discovered two areas in the left-brain hemisphere that indirectly supported Gall's theory. However, it was not based on bumps of the skull. It was a disease of the brain found after the patients were dead. It was not based on speculations, but on scientific cause and effect. Broca area is located on the frontal posterior lobe of the left hemisphere while Wernicke is in the left upper temporal lobe. Both Broca and Wernicke areas convinced mostly everybody that the left hemisphere was responsible for speech.

DR. PENFIELD AND H.M.

A scientist's goal trying to pinpoint a location for memory genesis and storage called the attention of the few and criticism of the many. Dualism was not completely dead. Karl Lashley working on rats seemed to bring down the theory of a specific location for brain processes. However, Broca and Wernicke's scientific discoveries did not go away. An enthusiast and brilliant researcher of specific brain functions, Wilder Penfield (1891- 1976) worked on awake and conscious patients under local anesthesia. He began to do his exemplary and exploratory brain research with epileptic patients. The brain has no pain receptors. When Painfield touched or stimulated different region in the temporal lobe; the patient, conscious and alert responded and reported what he had seen or feeling while under the eye of the surgeon. Remember, it is in the left and right temporal lobes that we have the hippocampus, a central center for learning and memory. I must alert you that the amygdala, center for fear, is in the central medial temporal lobe. Imagine Penfield's surprise when he touched a piece of brain tissue and his patient responded by telling him a past memory. As Penfield advanced his research, he stimulated different regions of the brain, taking notes of its functions. Patients were reporting memories from the past; they did not experience in the operating room. I want you to travel in mind and place yourself next to doctor Penfield and ask his permission along with the patient to touch a piece of his brain tissue. The patient may have said to you, "I cashed my check this morning before I came to this hospital." You had not asked him anything nor were you expecting him to tell you the whereabouts of his money. Won't you be surprised? I suppose that you and doctor Penfield will come out saying, Eureka, Eureka, I found it, I found it. Lo and behold, I have found the region for memory storage.

Doctor's Penfield discovery not only awakens the world, but it also aroused and sharpened the senses of many curious scientists

114

who turned their attention and tools to brain research. They were applying the most modern technology to map out the brain regions and its functions. The double helix and later on, the human genome, embryonic stem cells among others. Besides Penfield discovery, pushed psychiatry in the direction of modern science. Speculation gave way to laboratory work and research. Scientist around the world were reading and listening to Penfield's latest discoveries. However, despite his novel findings on memories, surgery on another epileptic patient known by the initials, H.M. placed everybody in suspense. H.M.'s hippocampus on both hemispheres had been removed. It had been done by another neuro-surgeon. H.M.'s epilepsy was significantly improved, but all memories after surgery never took hold on his brain. You could be talking with H.M. while having lunch together and shortly, after crossing the street he could not tell you what he had for lunch. Both hippocampi had been removed through surgery. It was another interruption in the brain research. Surgery on one side of the brain with epileptic patient had not provoked H.M.'s memory loss. It was the loss of both hippocampus and nearby tissue that left H.M. living in the past. He could not store new memories. The hippocampus was necessary for the storage of new experiences. Learning is based on memory and H.M. was unable to store anything new; he was left with what he had already learned despite his average intelligence. So, immediate or working memory for H.M. was out of his reach. Imagine if he were assaulted on the street, he would not be able to recognize the face of his assailant. It may sound like a joke to you, but imagine he had been married and his wife returned from the beauty parlor with a new wig and a new jacket. H.M. would ask his wife, who are you? Neuroscientist V.S. Ramachandran from California University had a similar case but from a different brain damage.

PHINEAS GAGE

There is another accidental case of a railroad construction worker that survived a large iron rod that went through his left frontal lobe exiting from top of his head and landing twenty seven yards away. His name was Phineas Gage. He was born in July 1823 in New Hampshire. He was working for a railroad company outside the town of Cavendish, Vermont. The accident took place on 9/13/1848; his death is assumed to have been in May 1860. After the accident, Phineas was able to talk and walk with very little assistance. He was seen by two medical doctors describing his injuries and behavior. The bed where he laid was bloody. He lost vision on his left eye, had a scar on the forefront, and on top of his head, there was a deep depression by which, you could perceive the brain pulsations, wrote the doctors. There are different interpretations of Phineas' behavior after the accident, but Dr. John Martyn Harlow, the physician who attended him during and after his recovery writes, the balance...between his intellectual faculties and animal propensities, seems to have been destroyed...gross profanity;... as a child he possessed a well-balanced mind...and was looked upon as shrewd, smart businessman...; In this regard, his mind was radically changed... his friends saying, he was no longer Phineas Gage. Precisely what was destroyed in Phineas Gage's brain was not known. We do not know whether it was only the temporal lobe or the frontal lobe was also injured in the accident. Some observations seem to implicate the amygdale and pre-frontal cortex; unfortunately, Phineas wandered about the Americas from the East Coast to California and Chile. No well documented records of his behavior were kept. Reproductions of his brain accident are just approximations of what might have happened, probably far removed from reality. In summary, we have covered the classical cases of Broca, Wernicke, Parkinson's, Alzheimer's, H.M. and Phineas Gage. In another essay, we covered the case of Terry Chiavo, Pope John II, Christopher Reeve and a little girl that at 18 months old began to lose some of her senses and ability to walk, secondary to lack of myelin, and the fibers that cover axons in the brain. We also covered in details a Parkinson's, Alzheimer's and a paranoid schizophrenic patient.

Under Luria's eye

Scientific research did not stop with H.M.'s accident, but stimulated young scientists to go for the biology of memory. Schizophrenia and Alzheimer's are two dreadful brain diseases in which, memory plays a significant role. Hallucinations and delusions are prominent in schizophrenia while memory loss is characteristic of Alzheimer's. In the latter, the patient losses his identity and personality; he becomes unable to care for himself during the advanced stages. In the former, long term medication may prevent suicidal attempts and hospitalizations in psychiatric hospitals. Alexander R. Luria, a Russian physician helped us identify regions and functions of the brain while he was attending wounded soldiers during WWI. Some of my colleagues wondered loudly how I allowed Mr. Delgo's memory or memories to remain in my brain for such a long period of time. I would like you to put yourself in my position. First, consider the strength of the fearful object. Mr. Delgo was the tallest man in the school; he was about 5 feet 9 inches and 185 pounds. I was a small and thin boy. At age 21, when I was inducted in the U.S. Army, I was 110 pounds. Adding to my nightmare, when I was fifteen years old, I fell in love with a girl in my classroom of about my age. Can you imagine how I felt when the teacher was punishing and torturing me in front of her? The idea of dropping out of school came to mind several times. However, my father used to tell my oldest brothers, "You do not give up at the first sign of pressure or pain. Stick to it and take it as a fruitful experience." One day I consulted the carpentry teacher regarding Mr. Delgo`s behavior and he asked me to talk with my father. Next, he said, "You know José, I see your father occasionally when I need his help. I could approach him if you want me to." I said, no, thank you. Following, he took me at the back of his store room and said, "Here you have lumber and tools your father has contributed to this program; you can come here anytime you want to. If there is anything else I can help you with, let me know." There were a couple of teachers I trusted and talked to them about my problem.

The girl I was courting once told me, "My oldest brother says he hates Mr. Delgo because he dropped out of school during the eight-grade because he used to pick on him every day. My brother once challenged him at the parents and teacher's meeting, but Mr. Delgo chicken out." After I graduated from College, I explored how parents felt about my tormenting teacher and two of them were very upset and angry at him. One could hardly contain his anger in that his face became reddish with his blood visibly enlarging his carotid arteries on both sides of his neck. He revealed to me that his youngest son had threatened to kill himself after he was verbally abused and tortured by Delgo. He had gone to Delgo and challenged him to come out his classroom and face him outside the school and students. He added that Delgo was a big bully inside the classroom but outside with men like him; he was a wicked coward. The school Director was an incompetent and drunken individual who depended on Delgo for leadership and discipline in the school. After this time, I dropped Mister from Delgo. Calling him Mister was a sign of respect as a professional person. From now on, he was neither respectful nor professional, but a wicked charlatan.

As I began to explore the biology of memory through the most recent researchers and their findings published in scientific journals and books, I learned that Delgo's toxic memories were interfering with my emotional well-being and learning processes. Recent studies with animals and humans indicate that chronic stress- the fear I experienced under Delgo- damages the amygdale and consequently, an adequate emotional response is seriously compromised, to say the least.

DRS. FRED H.GAGE & ROBERT SAPOLSKY ON CHRONIC STRESS

By association, I could have been in connecting learning environment, namely his classroom with subsequent future classroom, teachers and lessons with Delgo's noxious memories. Noxious memories would have sensitized existing synapses creating an emotional and learning block in my mind. Likewise, chronic stress and abuse has provoked neuronal death and synaptic freezing or contraction and inefficiency. The whole process of learning and memory in the amygdale and hippocampus in my brain was placed in jeopardy. Three years in constant fear is not a small time for a brain to become fractured or broken as has been called by neuroscientists. His noxious memories were occupying more than normal space in my brain. Professor Fred H. Gage at the University of California in San Diego wrote for Scientific American in 2003, "Chronic stress is believed to be the most causal factor in depression aside from a genetic predisposition...stress is known to restrict the number of newly generated neurons in the hippocampus."[36] Professor Robert Sapolsky of Stanford University writes for the same special issue of Scientific American as follows, "A stressor is anything in the environment that knocks the body out of homeostasis...; anxiety seems to wreak havoc in the limbic system...; one structure is primarily affected: the amygdale, which is involved in the perception of and response to fear-evoking stimuli.[37] The same article adds that prolonged exposure to stress hormone increases the risk of depression... and can lead to memory problems. Once more, my learning, my memory, mental and physical health was seriously compromised by Delgo's abusive behavior. Sometimes I wonder if my search for the soul more than compensated for a toxic memory like

---

[36] Fred H. Gage, Brain, Repair Yourself, Scientific American, 09/2003, p.51.
[37] Robert Sapolsky, Scientific American, 09/2003, p.87-91

Delgo. Learning for me must have been a rather difficult endeavor; I must have been in a continued high level of anxiety. My amygdale was in a high state of super-alertness. I did not give up. I tolerated abusive behavior but never lost hope. Ever since I remember, I have had in mind individuals like my father Polonio, Nelson Mandela and Mahatma Gandhi and recently, Luis I. Lula. President Mandela was tortured and incarcerated for many years, but never lost faith in himself as a whole person. He was released from jail and governed without vengeance to anyone. There seems to be an in-fight inside my brain between the soul and Delgo. Following my father's teaching, I kept repeating, keep on target, and do not give up. My father used to instruct me that if in case of danger I needed to retreat, not to hesitate as long as I took it as another learning experience. You do not always achieve victory by going forward, nor are the best routes in a straight line. I must have learned a lot from ideal objects for me to write this book at age eighty plus.

As I have stated earlier, J.P. Pavlov, John B Watson, Clark Hull and B.F. Skinner as well as Aaron T. Beck had come to my mind several times. It was not lack of interest in their theory and practice; rather I did not know how to put into practice their teachings with my patient's population. Theirs was a selective group of people with a specific set of symptoms. At the clinics that I worked for, we did not have that privilege; we had to serve patients with a myriad of mental and brain problems. From Eric Kandel, Antonio Damasio, Jean Piaget, J.Wolpe and a few others, I had read a few but very interesting articles. How to put into practice their teachings with schizophrenics, bipolar, Alzheimer's, depression, anxiety, and children with behavioral and learning problems was a huge challenge. It was not up to me to choose the patients that I could help the most with my body of knowledge and professional training.

DR. NANCY ANDREASEN

Never giving up the search for my brain functions, the time and person to rescue myself was soon to come. The year was 1984; the person was Dr. Nancy C. Andreasen and her book, The Broken Brain. It was written in a simple fashion for lay and medical students to find the most advanced biological psychiatric and psychological learning tools I could set my eyes and mind on. For me it was not only an eye opener, but brain busting. It was all there for me to enjoy. I said to myself, this book was sent to me from heaven if there is one. Having been blessed from heaven with Dr. Andresen's book, the following gift from heaven was, "The Brain, An Introduction to Neuroscience by Richard F. Thompson, 1985." Not long after Thompson, another intellectual jewel arrived from Gordon M. Shepherd.

With the Broken Brain book of Dr. Andreasen, my enthusiasm for neuro-anatomy, brain physiology and neurochemistry opened another route to the brain besides normal school of psychology. My College's basic courses in Biology and Chemistry in the fifties were rather rusty or very primitive. I needed to spend many nights and weekends making up for time lost in pursuing Delgo or perhaps, his counterpart, the soul. With Dr. Andreasen, I began to look at some of my patients as persons with medical diseases and emotional mal-adjustments in need of learning. I began to separate my patients into groups and develop specific teaching and therapeutic tools for each symptom and behavior. Besides my favorite tutor, Fritz Perls, I began to borrow from Behaviorism and many others culminating in an intelligent use of myself helping people to help themselves. Dr. Andreasen's book took me through a world tour on mental illness, I had never dreamed of. In Puerto Rico, I had seen schizophrenics of different types walking the streets while most people kept a good distance from each one of them. Most psychotic individuals were kept out in the country, cared for by distant relatives using homemade remedies. During my second year at the University of Puerto Rico, I volunteered to visit a grandson of mama Tiny who was hospitalized in a psychiatric hospital nearby the University. He was about six or seven years older than me; meaning I knew him from childhood. I suspect that antipsychotic medication was not available to that hospital yet. The year was 1956. I was informed that the patient was kept in isolation by staff members because he was very agitated. I looked at him through window bars. I called him by his name and he responded to me smiling and calling me by my name. He turned around and sat down on the floor against a wall. I never saw him ever again. The patients described in Dr. Andreasen book seemed to me to be different and in extreme cases of mental illness. In Europe during the Medieval Ages, the mentally ill person was considered evil possessed and often burned alive on a stake. A less cruel way of dealing with the mentally ill person was strapping him by his legs and tie him to a pole in a basement. Hidden away from family and public, he remained untreated and poorly fed. If the patient was accused of evil possession, the family and community members would consider his punishment justified and even call for his death. I came to the U.S.A. in 1949 and served in the Army from 1951- 1953. As a soldier, I had to be very "macho." There was no way you wanted to be seen by a psychiatrist. You did not want to be sent to the dungeons, called evil possessed and heavens knows what else. Psychoanalysis did not cure anyone that I know of; it was meant for mildly ill individuals with anxiety or mild depression. Dr. Andreasen's book and others opened the way for me to help myself and others. Almost torn into pieces, I still keep her book. From Dr. Richard F. Thompson's The Brain, I underlined in red the following: "The major damage in Parkinson's disease is the destruction of the dopamine's pathways from the substantia nigra to the basal ganglia. The basal ganglia are massively connected with the cerebral motor cortex…Parkinson's is a chronic and progressive disorder of movement…; the major symptom of Parkinson's is difficulty in starting and sustaining voluntary movements."[38] These symptoms were very close to me. My mother

---

[38] Richard F. Thompson, The Brain, An Introduction to Neuroscience, W.H.Freeman & Co. 1985, p.235,245

and two of my sisters were diagnosed with the Parkinson's disease. Very interesting to me at the time, and still is, dopamine involved in Parkinson's is also the neurotransmitter most responsible in another chronic brain disease, schizophrenia. Dr. Thompson wrote, "The extraordinary fact that all of the drugs effective in treating schizophrenia block the dopamine receptor, and are effective in proportion to how much they block it..."[39] Mostly developed in this observation, the theory of too much dopamine in the brain was attributed to the schizophrenic patient. To date, this theory has been modified to reflect modern psychiatry. I hope you can imagine how I felt at the time with a predisposition for thalassemia and Parkinson's. What should I do with dopamine, the pleasure chemical in my brain? Well, apply my teaching, keep on learning and thou shall overcome all problems.

---

[39] R.F. Thompson, op, cit p.321

NON-FREUDIAN TRAINED
PSYCHIATRIST

By this time, I was supervising psychotherapist at a community clinic, I was privileged to have an experienced psychiatrist helping me with medical and neurological problems. We used to discuss patients with both needs: psychotherapy and brain psychotropic drugs. At the time, I was working in two clinics. At one clinic the medical director was a non-Freudian European trained psychiatrist; we really enjoyed our brain and psychotherapy supervision. He introduced me to two professional magazines I still subscribe to. The number of subscription to professional magazines has multiplied several times, but Nature, Science News and Scientific American remain a strong source of reference for most of my intellectual curiosity. In fact, it was this psychiatrist who one day came with a box of about sixty copies of the above mentioned journals and said to me, "I have a gift for you, I selected it from my library; you will enjoy the articles I marked for you to read." I transferred the box to my car and I still keep most of it. We have come a far way in psychiatry and psychotherapy during the last 25 to 30 years. In my previous essay, I published my experience with Parkinson's, Alzheimer's and schizophrenia patients I have had the privilege of working with I tried to do my best to present those cases just as a brain disease or broken brain as Dr. Andreasen would call them. These patients are not possessed by evil spirits, and are not lazy individuals or faking their illness. In Parkinson's and Alzheimer's disease, there is a significant brain cell death. In schizophrenia, the latest discoveries have pointed out some genes responsible for certain symptoms of the disease. Most professional opinions claim there is a hereditary component in schizophrenia. However, there is not a single gene or group of genes that we can say that are strictly responsible for schizophrenia. We cannot say there is a single neurotransmitter for schizophrenia either. Furthermore, whether the dysfunctional schizophrenia pathway begins at the pre-synaptic or postsynaptic cell, this is still open for investigation. The over expression of dopamine D2 receptor is a serious suspect. There are so many symptoms in schizophrenia that makes it very difficult to say for a certain, this is the location in the brain we must look for. This brain disease exists in every part of the world. It does not matter in skin color, nationality, social class or the religion you practice. The best you can do is learning about its symptoms and take the person to a hospital or competent health professional for diagnosis and treatment. They will be treated humanely and prescribed the best medication in the market to control his/ her behavior, mood and thought disorders. Supportive psychotherapy heavily loaded with love, care, empathy and trust is essential in all cases.

All the cases we have approached in this essay have dealt with the brain. In all patients; memory, learning and painful experience have been the center of our interest. You may cut your finger, have surgery or fracture an arm or leg, and after it heals, in most cases, you will be fine. A memory scar is not subject to surgery, at least at present time. You do not see it, but for heaven's sake, you feel it day and night. You do not look for it nor want its company, but you cannot scoop it out with anything. You know it rests in your brain and you may even rule out regions in the brain for the location of your torturing memory. Medication may be helpful, but at present time those recurring and painful memories best respond to psychotherapy. In most cases, both medication and psychotherapy is best recommended. Most of my professional life I have dedicated it to work with memories, depression and anxieties. It is mostly logical that I dedicate the rest of this essay to patients I have worked with for the last thirty years. I want to point out a word or two of cautious. The euphoria partially created by the new technology brought about during the decade of 1990 and expounded by the media, claimed grandiose and unachievable goals in the short terms. Gene therapy, embryonic stem cells, organs regeneration and replacements as well as dramatic cognitive enhancement were mass production in the mind of science fiction. Somatic and brain cells are complicated, sophisticated and highly developed self-sustained chemical factory machines.

They began their present journey billions of years ago. They have had ample time to come to their current stage, our stage of development. Our brain now wants to accelerate and improve on what we have at the present time. There is no doubt we can do it, but it takes time. Hopefully, instead of weapons of mass destruction, we seem to be turning to tools for self-improvement.

THE D.N.A. AND
CATALYST ENZYMES

A brilliant researcher I enjoy reading during my spare time is James D. Watson, the co-discoverer together with Francis Crick of the double helix. He states, "All D.N.A. (my d. n. a., too) throughout the world is the same basic molecule, whether found in bacteria, viruses, plants or animals." Basic research in animals including humans has revealed part of the wisdom accumulated by our brain through thousands of years. There are enzymes-molecules- capable of provoking changes without suffering changes itself.

These enzymes can cut D.N.A., our genetic code, only when it recognizes a particular sequence in the D.N.A. It is known as restriction enzymes. The first restriction enzyme discovered was Escherichia coli, known as Eco.R1; a bacterium that reproduces and lives comfortably in the lower intestine; most are harmless but some could cause serious food poisoning. A second enzyme that chemically modifies D.N.A. sequences was soon found. Let's say that during the first restriction enzyme job, it may have cut D.N.A. sequence GAATTC. However, if the brain system for cutting D.N.A. sequences wants to bypass GAATTC, the second enzyme will add a methyl group, CH_3, to the bases GAATTC. This molecule is therefore modified and will pass unrecognized by Eco.R1.[40] This apparent simple process of sequence cutting and bypassing took our organism a long, long time. Two enthusiastic scientist curiosities over bacteria resistance to antibiotics, Stanley Cohen of Stanford University, a plasmid pioneer, and Herb Boyer, a star expert in restriction enzymes, became the world's first genetic engineers. Recombinant D.N.A. was born. This is not a small feat of science if you consider the time spent by your organism

---

[40] James D. Watson, D.N.A., the Secret of Life, Alfred A. Knopf, New York, 2006, p 89.

to develop it. It is another demonstration of the brain's memories to plan, organize, build and modify its own creation for its own benefit. You and I are benefiting from Boyer and Cohen's creative minds. Even better, most probably, they did not have Delgo in their brains.

There are countless dedicated scientists who work tirelessly to find the cause and cure for our illnesses. Lap-Chee Tsui and Francis Collins worked on the sequence of the D.N.A. and discovered that a patient with cystic fibrosis was missing a stretch of three base pairs, (like a-t, t-a, and g-c, for example, not the real ones) as a result of the absence of just <u>one</u> amino-acid in the protein.[41] The point that I am trying to bring out here is that after the discovery and construction of the double helix in 1953, research in biology and chemistry opened new horizons for American psychiatry and medicine in general. Brain diseases and disorders are subject to intense research at the best University laboratories and Research Institutes nationwide. It is not speculative work on introspective psychology but laboratory research done with animals including humans. Instead of the couch, tools like the PET scan, fMRI and in vivo clinical observations during hospitalizations have proven better psychiatry. Empirical psychology and laboratory biological psychiatry have joined hands or brains, for the good of the many. Pharmaceutical companies have contributed with medicine trials jointly with laboratory work.

---

[41] James D. Watson, op. cit. p. 310

ATOMS, MOLECUES AND BONDS

Responding to suggestions by friends and students, we are going to add a few basic – elementary facts about atoms, molecules and bonds before proceeding with proteins, learning and memories. From primary grades we have learned that all animals including the big whales, tiny bacteria, trees and vegetables that we eat in our daily diets are or come from living organisms. And organisms have different shapes responding to a genetic code located in the nucleus of the cell and input from outside. The inside of the cell is called nature and the outside is named nurture by lay people. The inside is what you bring with you from your parents and grandparents. The outside or nurture is the oxygen you are now breathing, the breakfast you had this morning and the ice cream you plan to enjoy this evening as a dessert. All living organism -life itself- had its origin on our planet earth around 31/2 billion years ago. How and where life appeared on earth for the first time is debatable not only among scientist, but also among theologians. However, considering that our planet was in a very stressful stage during early formation, whatever existed was subjected to multiple problems, risks and threats. Elements inside earth were taking their place in the core of the forming planet. Constant eruptions producing immense heat and toxic gases sprout into the earth's atmosphere threatening any form of life existing at the time. From space or from the sky, if you will, planet earth was bombarded by meteors of all sizes. Some scientists, based on well thought and elaborate theories, claim that there might have been a time when another smaller planet bumped into our earth and formed the ocean basis.

Despite of all threats and risks, a cell was formed, and in time, it could divide and multiply. Cells that did not divide and multiply are called unicellular organism like amoeba and the paramecium. You and I are multi-cellular organisms because our original cell was able to divide and multiply. Cell division and multiplication is a simple but complex process. There are cells with a nucleus known as eukaryotic cells and prokaryotic cells that lacks a nucleus. We are mentioning cells in this paragraph because our brain is composed of highly specialized cells called neurons. Your memories and my memories, the things that make you and me and what we really are, are stored in cells in our brains. This is not speculation but a scientific fact. From the nucleus of each cell emanates or originates signals for the formation of molecules ultimately forming proteins, the building blocks of our organism. In all types of organisms, there seems to be some type of developmental order that follows one after another. Looking outside, you will see a small plant. There is a time for that plant to grow from the ground, another for branches extending in many directions; there will be time for flowers, seeds and edible fruits. Plants and animals are made of the same basic elements and will grow, mature and die. There are 92 elements that appear on our planet in its natural or elementary form. Out of those 92 naturally occurring elements about 25 are considered essential for all forms of life, but four like carbon, oxygen, hydrogen and nitrogen make up 96% of all living matter. You add phosphorous, sulfur, calcium, potassium, and a tiny amount of few other elements account for most of the remaining 4% of an organism's weight.[42] From these 25 elements, our embryonic toti-potent cells formed or synthesized all types of specialized cells for our body. Lo and behold, because from the less abundant elements in our body, from the four percent, you and I will form the sodium and potassium pumps in the axon of our neurons in the brain. Those pumps will generate electricity. Without electricity generated in our brain, you and I will be extremely limited in every way. That brain of yours as well as mine, may be considered a machine dependent on chemicals and electricity to establish communication between its functional basic units, the neurons.

---

[42] Neil A, Campbell, Biology, Benjamin Cummings, California, 1993, p.26.

THE ATOM ACCELERATOR NEAR GENEVA, SWITZERLAND

Television programs present atoms, the basic components of any element, joining another atom by sharing electrons. When atoms share electrons, physicist calls it, covalent bonds. Besides covalent bonds, there are ionic bonds, molecular and hydrogen bonds. Double covalent bonds are all important in life. Some physicists have been able to break down the atom into many tiny parts often named waves. Most of the weight of an atom is occupied by the proton followed by the neutron while the electron has a very tiny relative weight. In CERN, Geneva, Switzerland, a group of nations built the most modern atom smasher, the Hadron Collider. The purpose is to split the atom nucleus and learn among others, content, behavior, function and possible usefulness in our technological world. As shown in children's books, all atoms of a specific element, for instance, carbon has the same number of protons in its nucleus. This number of protons in the nucleus is known as the atomic number. Some atoms have more neutrons than protons of the same element. Once physicist named these different atoms form containing more neutrons than protons as isotopes. Carbon dating is widely used to date on human fossils to understand human evolution from more primitive hominids like chimpanzee, Neanderthals, homo- erectus and Australopithecus. Carbon 12 and carbon 13 are stable isotopes, while carbon-isotope14- which have 8 neutrons in its nucleus is radioactive. An interesting note to remember is that our organism that was born to die, like all forms of life, was built on or around carbon atoms. All atoms of an element must maintain the same number of protons in its nucleus to keep the identity of that element. This is relevant, among other things, to learning and memory. Protein synthesis is equally important for good health. The shape of a protein determines its function. If amino acids are missing or a sequence in nucleic acids are mixed up, meaning altered, peptides, the proteins forming chain, would be seriously compromised and end up in a genetic disease. A note of clarification on atomic and molecular bonds. The bonds of atoms that will be forming molecules and cells are strong bonds; bonds for neurotransmitters across a synapse between presynaptic and postsynaptic membrane is a weaker type of bond. The receptor at the synaptic membrane is a protein that opens its door to a chemical with an anticipated or pre-ordained affinity. When a man and a woman marry, they have a matrimonial union, not a covalent bond. When two or more elements are combined together, it forms a compound which is neither a covalent bond nor a matrimonial union. When you have two atoms of hydrogen and one atom of oxygen, H2O or water, you have a compound. Compounds can take different molecules shapes. In a methane molecule, you have the carbon atom in the center and one hydrogen atom up, one down and one to the left and one to the right forming a diamond like shape. In the water molecule, the two hydrogen atoms seem to be hanging separately from the oxygen atom like a capital letter A, but without the horizontal line crossing it. Look for Campbell p. 34.

MEMORY AND ITS DEFINITION

The word memory comes from the Latin word, memoria. It has been defined differently by philosophers and psychologist in many places and time. In most primitive societies, memory is attributed to immaterial forces outside the human body. Most recently, we ascribe the brain the capacity to recognize previously learned behavior and experiences. Memories are stored in a variety of brain areas depending on the type of memory whether it is a working memory, long term memory or a visual memory. The accidental case of H.M. as we described in the previous pages shows to you how important is the hippocampus for immediate memory and storage. Visual memory with its complex pathway from the eyes to the occipital lobe at the back of the head was also compromised in H.M.'s When both hippocampus were removed, it took the capacity to store in other areas of the brain, the visual memory of the steak he had eaten with his friend the evening before. He would even deny he had ever gone to the restaurant or smelled and tasted a good wine, he had ordered. Had he gone to the Opera House to listen to either Pavarotti or Placido Domingo, he would argue with his friend that he had never seen or heard either tenor. In the case of H.M.'s behavior; visual, auditory, sensory and taste memories were most probably seriously curtailed or at least, detrimentally compromised. If the thalamus and pre-frontal cortex normally involved in immediate and long-term memory might have been left untouched in H.M.'s case, however, it could not help him because the important stop for storage and retrieval of memory had been removed. It might be a good argument for the brain as a unitary organ that receives, discriminates, analyzes, plans and execute functions and responses in consonance with all its component parts. It receives internal and external signals, creates patterns for action after it synchronizes all possible appropriate responses.

There has to be an appropriate response for each stimulus coming from primitive brain stem, emotional limbic system with its fast reaction amygdale and well-developed executive group of neurons, namely the pre-frontal cortex. Accordingly, there is a time to use a stick and fight, run or establish a dialogue. In the case of Phineas Gage, he lost the eyesight of his left eye; became ruthless and argumentative often provoking unnecessary aggressive and violent behavior, according to some authors. You can conclude that if the pre-frontal cortex was not directly damaged, its pathway connected to other behavior balancing organs was compromised. In both accidents, H.M and Phineas Gage were involuntary and innocent victims of outside intervention. In H.M.'s case, there was a physician attempt to improve his behavior while in P.G.'s; it seems an oversight on his part.

SYNAPSES

Most neuroscientists place memories at cells synapses, which are the communication space between neurons. From the very beginning with Delgo's memories, we have attempted to stay in the brain and neuronal communication. We did not explore somatic cells that compose our body and have their intrinsic memory for coding dividing and multiplying itself in many identical daughter cells. We pause with amazement and admiration for the working of life. Three days after a human egg is fertilized, it has been divided in eight cells; two days later, it already has about three times the number. In nine months, there are billions upon billions of cells forming an organism that in some cases will live to be in a hundred years old. Even better yet, in nine months, the original eight cells have divided to form sophisticated and specialized cells like neurons, heart, liver and eyes. As adult human beings, brain cells are already developing tools to repair bodily organs. Neuroscientists have undertaken the challenge to explore the brain and its neuron's synapses where the memory is to be formed and stored while there are trillions of synapses in the brain. How different memories are synthesized by the neuron itself and stored in different organ and brain tissue is another curiosity of mine that I need to explore. For the sake of curiosity, and perhaps a learning experience reinforcing existing synapses and creating new memories, we invite you to take a brief tour at an animal's cell, its main components and the inside working. We will not go into weight and size of the cell because just creating a mental image of around 900 billion cells-neurons and glia cells- in the brain are enough to give you an accurate impression of size and weight of each cell. Those 900 billion cells are inside your head, if you will.

We are going to describe you, a human cell, in particular, a brain cell. On around the center of the brain cell, you will find a nucleus contained by a membrane, and within the nucleus, you will see the nucleolus also surrounded by its own membrane that separates it from the cytoplasm. Outside the cell nucleus, you have the mitochondria, also contained in its own membrane. It is an energy producing organelle. The maternal D.N.A is found in the mitochondria. Following the mitochondria, you come across the Golgi apparatus. This organelle is involved in forming and packing multiple molecules, which will be sent out the cell besides guiding newly formed proteins to their destined compartments. The Golgi apparatus looks like flattened tiny plastic tubes that when called for, a small fragment will separate itself and deliver its cargo to inside or outside the cell. Adjacent to Golgi, but outside the nucleus, you will find another very interesting and fascinating organelle. It is called the endoplasmic reticulum. To us, this organelle looks like sausages in the super-market, but I call it the E.R. or emergency room because it is a processing and packing unit. In a hospital emergency room, patients are triage, seen or examined by physician and sent out home or admitted to the hospital. Once more, my memory from a previous job, made me associate both E.R. Unlike the Golgi apparatus, the endoplasmic reticulum is connected with the membrane of the nucleus membrane. We must understand the cell has survived billions of years and had developed very sophisticated mechanism and tools for the manufacture of necessary substance for its own use. Inside the cell you have organelles and microtubules performing a myriad of functions without interfering with the specific function of each cell. There is order and a hierarchy of jobs to do for each individual component of the cell that has guaranteed its efficacy through evolution.

Besides organelles for processing, packing and exporting substance from the cell, you will find an important organelle dubbed lysosome. Cells capture material from the outside the cell medium and is brought in to the lysosome for cellular digestion. Actually, when the material is captured by the membrane plasma, a tiny portion in the shape of a sphere, splits off and fuses with lysosomes for intracellular digestion. There are continuous interactions among organelles inside the cell. For better understanding, you can form a mental image of the interaction between lysosomes, endoplasmic reticulum, and Golgi apparatuses as outside substance that is brought in for digestion, processing and export. In the cell body, you will see the mitochondrion providing all the necessary energy to maintain the cell multiple processes going on. Besides the organelles that we mentioned, there are vesicles inside the cell traveling from one end to another performing all types of cell and life functions. Within the endoplasmic reticulum, where a lot of processing and packing is taking place (we are unconscious of it), you will come across many tiny spots. These tiny molecules go by the name of ribosome. They consist of ribonucleic acid and proteins. Ribosome has been likened to assembly factories. This is the location where multiple proteins are formed or synthesized. A single strand of D.N.A. named messenger R.N.A. carries the message-nucleic acids code- for different proteins. The molecule messenger R.N. A. brings the code and another molecule, transfer R.N.A., meet each other at the ribosome and join to form peptide lines. Proteins are peptide lines.

The accepted dogma for protein synthesis is D.N.A.—R.N.A. proteins. This dogma was thought of by Francis Crick while he was working with James Watson on the D.N.A. double helix. The enzyme R.N.A. polymerase catalysis the production of a single- stranded R.N.A. chains from double-stranded D.N.A. template.[43] The process of protein synthesis was described to you in above pages, but we will continue to elaborate a little bit more for better clarification and understanding of this simple but complex cellular process.

---

[43] James D. Watson, op. cit, p.69.

CHROMOSOMES

The complex (combination, structure) of D.N.A. and proteins that form a chromosome is called chromatin. As the chromosome prepares to enter mitosis (cell division), the chromatin thread or string coils up forming a solid structure. This solid structure must take place because chromatin exists in different states at different phases of the cell's life cycle.[44] Chromosomes need to be a highly compact and condensed structure to stand cell division. Otherwise, the chromosome would be subjected to external threats and invasions by extra-cellular polymers existing in the surrounding medium. The chromosome D.N.A. needs to be well protected from internal as well as outside intervention. It must maintain a strong structure to survive cell division and still, conserve its internal genetic code. Changes in nucleic acids- changes in A-T, C-G sequence would not only create mutations, but life threatening diseases. "A single base change in the D.N.A. sequence of the human beta hemoglobin gene results in the incorporation of the amino acid valine rather than glutamic acid into the protein. This single difference causes sickle- cell disease…"[45] There is multiple diseases product of changes in nucleic acids sequence as well as molecular failures. The bases in the D.N.A., our human genetic code has survived millions of years maintaining its basic structure. There have been slight changes due to minor internal and external changes, but the basic structure remains unaltered.

Amino acids are the building blocks of proteins and proteins are the building blocks of our body. Three nucleotides code for a single amino acid. We can say that a codon (three nucleic acids) code for a single amino acid, which will form peptide chains, which are proteins. Differences in the sequence of nucleic acids adenine, thymine, cytosine and guanine-A-T, C-G, for short, in the D.N.A. code is

---

44 Alberts et al., op. cit. p. 250.
45 James Watson op. cit. p. 67.

traced back to differences in the amino acids of proteins. There are twenty amino acids for proteins synthesis and three nucleotides-3 letters- for each amino acid, meaning that three bases or a codon, is necessary for each amino acid. It took time and hard work for chemists and biologists to discover that three-letter code for each amino acid. In most cases, the combination of three-letter code for more than one amino acid. For example, in the amino acid alanine, we have the codon ACA, ACC, ACA and ACU. All four different codons code for the same amino acid. In the last codon, also known as a triplet, you see a u for uracil because in the R.N.A. chain, uracil replaces thymine. The difference between a D.N.A. and an R.N.A. strand is the introduction of uracil in place of thymine. Both, thymine and uracil bases are complimentary to the nucleic acid base, adenine.

THE BRAIN QUAGMIRE

Trying to escape from Delgo`s toxic memory while I was interested in the "niche" or site of the soul in our brain, I became trapped in a brain quagmire. The quagmire that I fell into is: when and why does the brain decide to convert noxious short-term memories into permanent ones. Short-term memories like working memories do not last long in the brain and generally, do not hurt unless it is converted into long-term memory. It seems to me that it is a paradox. The evolution of our brain has overcome many risks and life-threatening episodes and, can now even go into a deeper phase of development by self-repair. On the other hand, why would this powerful organ create a memory that could destroy itself? We have come across soldiers from WWII, Korea, Vietnam as well as Holocaust victims in need of intensive psychiatric and psychological intervention. Furthermore, during interviews with their relatives, we have learned that some of them committed suicide. The point is that the brain synthesizes memory molecules that end up destroying its own creator, the brain and the whole organism. We are aware of apoptosis (programmed cell death) and cancer cells as well as genetic failures or mistakes. However, these are different processes from self- destroying memory molecules. It is widely known that protein synthesis is absolutely necessary for associative learning to be integrated into long-term memory. Eric Kandel, Nobel Prize Laureate and well-known researcher on memory states, "Long-term memory requires more persistent synthesis changes that are based on alterations in gene expressions as well."[46] My own noxious experience, I believe may justify the changes Dr. Kandel refers to, in the above quotation. I was subjected to torture for one and half hour, five days a week for three years. I repeat, it may meet the requirements for long–term memory and alterations in gene expression. My brain was receiving sight, hearing, and with your per-

---

[46] Eric Kandell, op. cit. p.292.

mission, skin stimuli just by the physical proximity of Delgo's huge and obnoxious body. Most importantly, it was not only 1 1/2 hour a day, five days a week; the psychic or brain reality of torture was 24 hours a day, seven days a week for many years after I left school. Eric Kandell wrote in the same book that short-term memory provides a change in the function of the synapse...; long- term memory requires anatomical changes. Dr. Kandel's discoveries have been replicated around the world. Neuroscientist at the Max Plank Institute in Germany, the Unites States and Japan have confirmed Kandel`s original findings. Dr. Jeanette Norden, Professor of Neurosciences in the College of Arts and Sciences at Vanderbilt University states, "Many of the processes that are involved in learning and memory in humans are conserved throughout the animal kingdom." I do not need to remind you that our human species is part of the animal kingdom. She added, "The sea slug has many things operating that allow him to learn, and the basic mechanisms that allow learning in this animal are the same mechanisms that take place in our brains." In another paragraph in the same page she wrote, "We think that long-term memory involves some kind of change, actually, of proteins synthesis in neurons in widespread areas of the cortex."[47]

---

[47] Jeanette Norden, Vanderbilt School of Medicine, The Teaching Co. Part 3 0f 3 Virginia, 2007, p. 27

CLARIFICATION

A note of clarification to our readers in relation to our professional and clinical experience with patients we have come in contact with. Despite the humiliation and pain that we went through Delgo, we dare not to compare our experience with that of warzone veterans or Holocaust victims. It would be unforgivable and sinful if we ever dared to connect one with another. Of course, on different occasions, I wished I were dead, but there were no bullets or crematory chambers waiting for me. It was psychic pain alright, but I had the safety of my home, family and friends to go to. There were lapses of time when I took the position of; "I do not care anymore." It is a bad situation to be in; depression begins to gain access to your brain, into your entire being. Your brain chemistry is altered and neurons synapses become much less active, meaning, neuron firing is rather abnormal. It is like life has no meaning for you. The joy of living has stopped and nature beauty does not exist for you anymore. There is no time even for day-dreaming. The brain seemed to produce one thought only and that was Delgo. You might rightfully wonder whether I was ever hospitalized for a psychiatric problem. Fortunately, the answer is a no. More for curiosity than necessity, I visited three different psychiatrists and got two different diagnoses. Never took prescribed medication from a psychiatrist. However, I have been in therapy, both individual and in groups for years. When I was going for psychoanalytic training, I had to go into therapy as part of my training. Besides my training as a Gestalt psychotherapist and a group training, I spent two years on the couch. I learned a lot about myself that otherwise, I would not have learned. Besides, I took vitamins and minerals as recommended by experts in the field.

MY SISTER, DESDE

A few pages back, I said my next older brother, Jack died of what seems to me now as thalassemia. I have thalassemia traits, as well as two of my sisters and a brother. He died before I reached my first birthday. My mother entrusted my care to my eldest sister, Desde. She took care of me until I was about six years old. Desde got married and left for the nearest town with her husband. I felt quite bad and cried for her return home. She used to come once a month on Sundays. We do not need to tell you how happy I felt when she showed up out in the horizon coming home on Sundays. During therapy, I had hardly any memories of my mother as a little child. All my childhood memories were centered on my sister, Desde. I enjoyed when she cooked oatmeal and served my plate in a corner of the kitchen. We lived in a two-story high building with a huge kitchen. The first floor was a storage house for maize, beans, tobacco and sweet potatoes. Those were my father's favorite crops. The kitchen was located on the second floor followed by the dining room, but my sister preferred to serve me alone while she watched over me. She always demanded that I eat everything on the plate and I was more than happy to please her. I tried hard to remember my mother during this period of my life, but she did not show up in my brain. My mother had been an orphan at age five; her father remarried and she worked hard at home. She did not seem affectionate with her children. In short, my mother was not emotionally there for me during my early years. My recollection was anger towards her. When Desde left my home after her marriage, I remember trying to do things to gain the attention of my mother, but she did not respond to it. She had two younger children to take care. I remember bringing her candies or other minor things on mother's day, but it did not produce the kind of response I was expecting to receive. Even more painful were memories when I used to save pennies my sister used to give me and I saved it to buy my mother a present on mother's day. I was trying to buy my mother's love, but it did not work. In retrospect, my sister and mama Tiny made for her emotional vacuum. Also, my therapist was an intelligent and concern lady who guided me very well. She helped me go deep into myself and draw from the wisdom I had gained through my neighborhood experience. She believed then, as I do now, that we can build new positive experiences to replace old and sour ones.

MY MENTAL LIGHTHOUSE

Half of my life span was spent dealing with the scar left by Delgo. I said that I was dealing with, not that I destroyed or eliminated Delgo from my brain. What I suspect I have done with Delgo is create new synapses or perhaps new dendrite buds partially isolating the old ones. Now we turn to my mentor at a distance; I say at a distance because I have never seen her personally. Her books have been my mental lighthouse for about 25 years ever since I read her book in 1984, The Broken Brain. She is Dr. Nancy C. Andreasen, Director of the Mental Health Clinical Research Center at the University Of Iowa Carver College Of Medicine. She has written several books besides numerous scientific articles published on professional journals. In her book, The Creative Brain, 2006, there is a section on memory, specifically, episodic memory that relates to my experience. Episodic memory is defined "as autobiographic memory, the recollection of information that is linked to an individual's personal experience…The capacity to place events in time and to reference them to <u>self</u> may form the basis for self- awareness or consciousness."[48] There is no doubt in my brain-mind- that Delgo falls into an episodic memory. In my case, it took place between the ages of 13 to 16 years old. It was a school in the countryside. At the time there was no state or local government law that required parents to send their children to school beyond sixth-grade. It was a public and free educational system during the primary level. Besides free elementary education, there was a free lunch for every child attending school. It was a very good incentive for poor families to send their children to school. At home, there was an abundance of fruits and vegetables; we never went to bed hungry.

Following Dr, Andreasen on episodic memory, she adds, "Episodic memory permits the individual to reference his or her

---

[48] Nancy C. Andreasen, the Creative Brain, Plume Books, New York, 2006. p.71.

personal experience in both time and space; as such it creates the experience of consciousness, the sense of individual identity, the ability to refer one's own experience to that of others, and the capacity for introspection.[49] In reference to time and space, I already covered it in previous pages. In terms of consciousness and self-identity, it has been my personal struggle ever since I began to feel that I was closer to my sister than my mother. I began to look at myself as different from everybody else around me. Perhaps the worst part of my life experience besides the emotional scar created by Delgo and my mother's emotional distance was the feeling of self-limiting, of not bringing out all my potential. It is a voice- like telling me to stay in the background; I do not deserve to enjoy myself. Of course, self- teaching during self-analysis following K. Horney and other psychologist and above all, formal psychotherapy begins the process of awareness. Consequently, the person is more able to participate in pleasure-producing activities, and begins to be able to make rational decisions. The process of choosing rationally takes over from prior behavior of accepting the unconscious dictate of self-depreciation imposed on you by a toxic internalized object. Rather than regressing to primitive instincts, the person can face crises in his life intelligently and creatively. Early in life, during our late teens and early twenties, we began making an inventory of our negative and positive attributes as well as individual person's actions that might contributed to our precarious predicament. The revival of bad experiences serve to allow us to feel the intensive pain while conscious, and then, reject and destroy the power we had attributed to the toxic object. The process of revival, rejection and destruction of the toxic object must be done with intense emotional commitment. It must be a decision making, focused and consciously directed to the target we want to destroy. Forgive me for repeating it again; it cannot be an intellectual exercise. You must put your chemistry, your emotions- to work for you. You can rest assure that when I was in front of the class accused of being poorly dressed and ordered by Delgo to kneel down on sand on a wooden box, my brain was running deep in its chemistry. As an

---

[49] Ibid.

emotional organism, neurotransmitters like glutamate, GABA, sero-tonin, acetylcholine and more were running high making circuits and pathways needed to address the recovery of good health. This initial process of self –contact where physical sensations and feelings are explored and expressed, builds a foundation for continued explo-ration, revelation and self- realization.

PEDALING DID NOT COME OUT OF THE BLUE

Pedaling, the object Delgo that did not come out of the blue; our brain has multiple solutions to survive challenges. Through, its trillions of synapses the brain is busy working out patterns and designs to solve conflicting input and bad teaching experience. Unconsciously, our brain is always turned on tending the needs of its own carrier, our body. The brain has to provide for our heart to send in blood with oxygen and sugar as well as carrying it to everywhere it must go for survival. In retrospect, the environment surrounding my childhood was not welcoming me into the world. Despite all types of abuse and neglect, my brain was in high gear defending itself from outside attacks. My brain was filling in emotional lacunae and physical abuses with the best nature had to offer... In psychoanalytic parlance it states, "The parts of the psyche (I call it the brain), which have to do with exploiting the environment gradually develop into what we call the ego. Consequently, the ego is that part of the psyche (brain) which is concerned with the environment for the purpose of achieving a maximum of gratification." [50] The brain at synaptic level was receiving all types of stimuli and assessing in a synchronized fashion its needs to be able grow naturally and healthy. Like it had done for millions of years, our brain went to places where it could get at least, the basic needs to continue growing. And, my brain did remarkably well. I do not think you need more proof that at eighty years of age, I am writing to you this long, but enjoyable life experience. When we learn to trust our brains and listen to its messages, the book of our life is well written. Someone said long time ago, the wisdom of the universe lies inside yourself. We read books to learn from others, but

---

[50] Charles Brenner, An elementary Textbook of Psychoanalysis, Anchor Books, New York, 1974, p.37.

it is the brain capacity to form patterns and designs of what we read to make it as permanent knowledge for use when needed.

To satisfy the curiosity of our students, we are returning to Dr. Kandel for a moment as it relates to permanent anatomical changes in our brain, when we form long-term memory like in episodic memory. Dr, Nancy Adreasen's writing on the subject states, "Both kinds (short and long-term memory) are due to plasticity in the synapse, which changes in response to stimulation and experience. Long-term memory occurs when a stimulus is intense and long lasting, prompting the nerve cell to send a message to the cell nucleus that leads to the synthesis of proteins that in turn produce long lasting changes in the shape of the synapse. Through Kandel`s work, we now understand the molecular mechanism of consolidation and storage." [51] Professor Norman Doidge at Columbia University States; Eric Kandel was the first to demonstrate that when we form long-term memories, neurons change their anatomical shape and increase the number of synaptic connections they have to other neurons.[52] Ever since I became familiar with Watson and Crick mind bugging discoveries followed by the double helix chain-based pairs, I looked at the cell nucleus as being responsible for my own existence as a human being. I said to myself in solitude; even the messenger R.N.A. is just a piece of the nucleus, a single strand of the double helix. Trap and unravel the message the nucleus sends out and you will come across the beginning of life. It was a naïve fantasy, but not far removed from the real thing. In May 2010, pioneer scientist, Craig Venter announced to the world that his laboratory had created a synthetic bacterium.

---

[51] Nancy C.andreasen, Brave New Brain, Oxford University Press, Oxford, 2001, p.63.

[52] Norman Doidge op. cit. p. 218.

INSPIRATIONAL & EDUCATIONAL PERMANENT MEMORIES

In addition to Dr. Kandel`s discovery on long-term memory and anatomical changes I quoted to you, I added the support of two additional neuroscientists, Dr. Jeanette Norden and Dr. Nancy C.Andreasen. They are researchers and professors at their respective Universities. In addition to episodic, semantic, working and associative memory, among others, I like to add another type of memory that unlike a toxic or noxious memory has been in my brain for a long time. This memory has been my inspiration in writing, as well as cognitive development and life enjoyment ever since I first discovered it in the summer of 1956 while attending University of Puerto Rico. It took place between 12:30P.M. and 1:00 P.M. I had classes at 1:30 P.M. After I had lunch, I went to my room to pick up my books and head back to school. For unknown reason, I laid down on my bed (not sure I did) and I felt I was transported out of my room into space. It was not a mind-travel as I have written in previous books, nor I was daydreaming. I was conscious I was traveling, and traveling very fast that I could not or did not have time to argue and questioned why and where was I taken. It is like I was overwhelmed by their power that I did not put any resistance. I said "they" but in reality, I could not say what it was. Before I realize it, I was in a cylindrical energy capsule out in space communicating with other energy cylindrical capsules by what I believe was telepathy. We did not have human or animal shape of any kind, we moved around in what supposed to be a huge spherical-like capsule. No food or drink was served. They were lecturing me in a very polite and sensible fashion, the nature of life on planet earth. I became aware of my essence as an energy capsule separate and distinct from life on earth. During the process of their teaching, they made me become conscious of the life form as animals and life form as they existed and experienced it.

After the teaching, the lesson was over, I was informed that I had to return to earth, to my life form as an animal, as a human being. I begged them not to send me back; that it would be an enormous punishment for me to return to crude animal form. It would be horrible and painful to exist in human form. I begged them over and over again to allow me to stay with them, but the last word I remember from them is the following, "It is not time for you yet. You must return to earth."

Back in my room from that unexpected trip, I ran out to the University campus. A friend of mine who saw me first asked, what in heavens have you seen, Jose? I smiled and kept on walking to my classroom. A few days later, I wrote down in my notebook that it was an educational episodic memory provoked by involuntary traveling in space and time. Involuntary traveling is not an original idea. C. Jung deals with it as well as Hindu literature, which were available to me through my brother's books. Actually, in my book, Puerto Rican Spiritism, Religion and Psychotherapy indirectly deal with the subject of involuntary traveling.

This educational episodic memory has been one of my companions for 54 years up to date 2010. It has motivated me to explore and study my own origin, my emotional and cognitive components, and most of all, my brain as a source and producer of electricity. During my educational traveling, I was made aware of my existence as a tiny capsule of energy, never created nor destroyed, but transformable. Nine years later, while I was living in 94th street and West End Avenue on the Westside of Manhattan, I had another educational episodic memory, an involuntary traveling in space and time. It was past 9:00 P.M., I was reading the works of Sigmund Freud. I do not remember whose translation it was. The door to my room was locked. Out of nowhere a medium size lady appeared in my room and laid down to my side, on the right side of my bed. I weighed 120 pounds at the time and the bed kind of bent towards her side. It was something with weight, but invisible to me. I became extremely scared of the lady, if it was a lady. I turned over on my left side and away from her and with my right hand, I held tight to the bed's horizontal railing. I was taken away against my will. I was extremely frightened, I never saw the lady again. Initially, I was placed in a cylinder with rings around it. I was traveling head first at a very fast speed. I had a feeling that I was going to get hurt, but suddenly I came out into another huge sphere or perhaps, another level of life existence. Like nine years before, I was in an energy capsule like the rest of them. We communicated through telepathy because we did not make any sound nor we had any animal or human body. Unlike the first time, I was asking all types of questions, including specific questions to each one I met. I asked each individual, meaning capsule, what specific role and responsibility each one had in the place we were in. After a short while of questioning, I could sense they were unhappy with my questioning and took me before the presence of around 5 1/2 feet tall, 18 to 20 inches wide light- energy- figure. The light, as I approached it, was a blinding clear light, but as soon as I came close to it, I was no longer blinded by it. I did not speak or communicate with anyone during my encounter with this light figure, but I was sent back to earth. Like the first encounter, the purpose was teaching me about the form of life on earth and our life experience. I did not feel threatened by them but I was very curious about their ability to exist in energy form. I was aware of my limitations because they had brought me in, against my will. I remember asking for some names, but I could never explain nor remember why I was asking for anyone. I want to emphasize that during my second involuntary traveling I was not a passive guest. I wanted to know the source of their power. I have had the feeling that my questioning was the reason I was sent out back to my room at an unbearable speed that I thought I was going to crash and die when I hit the building I lived in. That was the last of two traveling experience I had at that address. There were no Star Wars films or Star Trek television episodes yet for me to associate it with.

Two Educational Episodic Memories

The first educational episodic memory at the same building had taken place a few months before early 1965. I was not taken away; but it was a self-revealing and excellent teaching lesson.

To consolidate my educational episodic memory from short-term memory to long-term memory, it required anatomical dendrites or spines changes, which in turn would create new synapses. These are not noxious or painful memories but stimulating, positive and supportive episodes that guided me in pursuit of self-realization and happiness. Educationally, this traveling experience helped us get a better understanding of our own potential and limitations. Theories of multiple universes, quantum theory, black holes, negative and positive forces which, in reality are complementary forces are memories humans have discovered relative recently in time concept. These are memories created into proteins by the nucleus of the neurons under an electrical- chemical impulse or message. As you are already aware through previous pages, if there is no electro-genesis, there is no memory; there is no storage and there is no synapse formation or excitation. The most tragic case that we have shown to you is a person with the Alzheimer's disease. During the advanced phase, the patient with Alzheimer's is not aware that he/she exists as a human being. You can see her lying down in bed but not responding to any type of stimuli. She would defecate and urinate in bed without ever knowing that she is naked and soiled in bed. She would not respond to her name, thirst or hunger. If the brain has not accumulated-stored- memories, there is not a person to be identified with. Your family pet, be it a cat or a dog, respond to your verbal and emotional messages, but when neurons in a human being have collapsed, there is absolutely no response to stimuli.

TRAVELING EDUCATIONAL MEMORIES CONTINUED

My traveling memories, that I dubbed educational episodic memories, E.E.

M., for short, are subject to multiple interpretations by Freudian and Jungian psychoanalysts. However, if we put aside these two great psychologists and continue to concentrate on neuroscience, we may be able to learn something from the traveling brain. In total, I have had four traveling episodic memories that I consider helpful and contributed to my own self-understanding and self-identity. It must have been during the year 1960-61 that I had this educational experience. It was long before the first Star Trek television show or Star Wars movies that came along afterwards. During this E.E. M., I did not travel outside my body. Rather, I was taken on a tour through the planet earth in time and space. We started the trip in the utmost western tip of Europe, traveled east passing through France in the direction of southern Russia. We must have stopped in the Black sea, crossed the mountainous Caucasus and continue east to the Caspian Sea, and must have visited a primitive group of people along the Volga River. We also visited shanty towns in the Mesopotamia region and turned back to what is known today as Asia Minor. We proceeded south to Egypt, where we found also a primitive civilization, which my companions took upon themselves to educate for future generations to come. In May 2010, to my surprise and delight, my eldest son Joe, showed me a movie that depicted the same trip I had almost fifty years ago.

My interpretation of multiple individuals in different places having or sharing the same brain experience is that there must be a group of neurons forming an almost exclusive brain cells network or module that are able to form memories related to space and time. There are neuronal modules for gifted children that at a very early age, show their unique ability in music, arts, mathematics, physics, etc. In some of these cases, their parents are illiterate individuals with no chance of teaching their prodigious child anything in the subject they master like nobody else ever had. Those brain circuits are very well wired from birth and only need a little incentive or the material for them to create for the outside world the object that resides inside their brain. Specialized brain cells may only need simple stimuli from another group of brain cells or an outside stimulus to create the unthinkable. Neuroscientist, Michael Gazzaniga wrote, "Modularities are highly specific. Time, space, affect, and a host of other dimensions of a stimulus event all have to be encoded, and apparently are encoded, in different brain areas." [53] Work done in animals, including humans, have demonstrated over and over again that different regions of the brain, as well as neuronal groups, are specialized to store and respond to specific stimuli. For example, for most people, the left hemisphere is dominant for language. Damage to the Broca area as well as the Wernicke area will prevent the patient from forming words and sentences to verbally express himself. If it is the Wernicke area that is damaged, the patient's verbal production is incoherent. During brain surgery while the patient is awake and alert, if the surgeons touches or stimulates with an electrode a group of brain cells, the patient may respond saying, "Oh, yes that is the apple I had for supper last night." In reality, the surgeon was not showing his patient an apple, but the surgeon stimulated an area where that memory was stored. Some patients with brain anomalies when asked to draw a picture of, let's say a person, some draw only half the face and being convinced that it is the whole face. Loss of memory can be temporary in some cases.

---

[53] Michael Gazzaniga, the Social Brain, Basic Books, New York, 1985, p. 103.

OH, MY GOODNESS
FROM JANE DOE

While I was working at a Hospital Psychiatric Emergency Room, I was assigned to interview a lady (Jane Doe) of about forty years old. She was brought in by the local police. She did not have any personal identification; she did not know who she was or how she got there. She looked very sad and anxious. When I tried to get an information from her, she brought her hands up over head and looked somewhat upset while I kept asking her questions. After about 1-1/2 hour, I left my office and went to get something to eat. A lady nurse assistance kept an eye on her while I was out. Upon returning to the office with my food, clumsily I placed my plate on the table and a medium size decaffeinated coffee on the side. I signaled the lady if she wanted some of my food, but she shook her head. When I tried to take a bite of the chicken leg, the plate and the coffee got all messed up and fell to the floor. I had made a big mess of food and coffee on the floor that must have made an unexpected emotional impression on the lady. Her first reaction seemed to be surprised and bewildered at my shaking hands, head and clumsy behavior. A few seconds later, she was smiling followed by taking her hands to her mouth ultimately saying, "Oh, my goodness." Next to my plate I had placed my car keys together with my home keys. It had fallen to the floor next to the plate. When I bent down to retrieve my set of keys, the lady kind of shouted, "My car keys, where are they?" After hearing her unexpected shouting, I left my messed-up office and both of us went to my friend psychiatrist in his office, three doors down. We learned from Jane Doe that she was out of town going through a divorce process and had a very bad argument with her son. She was able to tell us that she had been driving for many hours. She remembered her car's plate number as well as her car's maker and model. The police were notified and later found her car near the place they had picked her up, aimlessly wandering around the neighborhood.

I have attempted to understand the mind and the brain as a single unit. As far as I understand it now, the mind is a group of neurons interacting with each other in very peculiar fashion. Together, it forms patterns, designs, plans and execute actions independent of anyone's consciousness. Only the most explicit and concrete behavior of our unconscious physiology becomes aware to us. For example, I will experience excruciating pain while attempting to walk if cells in the substantia nigra compacta died out and cannot produce the chemical dopamine to be sent to another group of brain cells known as putamen and caudate. Just a relative tiny portion of the total brain becomes dysfunctional, and I will be trying to hop to stand up and walk like in Parkinson's disease. Walking, like flying is an evolutionary process brought about by our organism over millions of years. In most animals, including birds, it is an inherited instinct. You can watch eagles trying to fly as soon as they have feathers. Likewise, you can see a small giraffe trying to get up and walk just a few minutes after birth. A human baby takes a lot longer to learn to walk. Walking, jumping and running are a big step forward if we consider that a few millions of years ago, we were traveling the savannah on four. Our ability to fly, meaning fly in airplanes is the product of accumulated or stored memories and retrieval.

Memory formation and the story of Brandon

Brandon is a two-year-old boy who walked and began to talk at normal age. He is very attentive to children's television programs. He tries to mimic T.V. sounds and gestures often to the detriment to those around him. He also likes to play with the dog of the house, a Fox Terrier named Nimov. Originally, Nimov enjoyed his games, but lately he tries to avoid Brandon and his overly aggressive behavior towards him. For some of you, we become obnoxious repeating just a simple task of fitting the left shoe to the left foot, and not in the right, except when intoxicated, is a synaptic process. We watched our youngest family member, Brandon, just two years old play with geometric forms and selective alphabet letters brought in for learning and memory forming. It was the first time we introduced toys to him. After a couple of trials, he learned to place 20 geometrics forms and letters in its appropriate place or boxes in the wooden plane. What surprised us most, was that he had learned in a very short period of time, the shape, size, color and box position for each letter and geometrical design relative to each other. For example, the letter T with a normal design as opposed to a T up-side down. The letter Q in normal position with an up-side down Q position, was easily recognized by him. Similarly, a rectangle placed in different positions was also quickly recognized by him. We threw into his working table another puzzle, letters and geometrical forms of different colors. He immediately pushed it aside; we tried smaller letters and geometric figures but Brandon pushed it aside, too. We could not explain how it was possible for him to learn size, shape, color and position of each letter and geo forms in a relative short period of time. The designs of working memory in his synapses and the designs in the working plane on the table were complimenting each other in an incredible very short period of time. For example, the letter T and letter L could be placed in four different positions each and he would find the correct place or receptor in the working plane on his table. In more than one occasion, he struck the box with non-fitting letters and geo forms to the side, as if letting us know that he knew in advance it will not fit. For the next step in our puzzle solving game, we provided Brandon with three and four-letter words for him to find the appropriate receptor. Besides this memory forming exercise, we introduced names of objects such as dog, cat, head and home.

Our curiosity in Brandon's experience was the issue of working memory, short-term memory and long-term memory formation. He has done the same game over until he lost interest for it. We had to bring more challenging games for him. Obviously, Brandon was enjoying working with the puzzles. Synapses in particular brain cells became activated, meaning more excited than normal. We can say those synapses were sensitized to that type of learning and pleasure producing games. We are assuming that those sensitized neuronal axons and its corresponding synapses would easily activate again with similar learning experience. Someone said, "Neurons that fire together stay together." It has been proven over and over again by researchers in brain plasticity. In this case scenario, learning is a voluntary and pleasant experience. There was no teacher Delgo for torturing. Brandon preferred playing with geo forms and letters because it provided him a pleasurable feeling. It means, the internal physiological/chemical rewards he receives each time he played the game. We assume that Behaviorism would consider it part of its learning theory. It is not pretentious to assume that introspective and Gestalt psychology can claim it for itself based on the premise that an organism tends to repeat an action that produces pleasure, reduces tension and ultimately, improves itself. With Brandon, he did not only receive pleasure through the release of dopamine and endorphins, but from our continued support, care and love. We hugged him, kissed him and reinforced his behavior each time he achieved his goal. There is no doubt in our brain-mind- that permanent synaptic structural formations are taking place in Brandon's brain each time we play with him. We provided him with an enriched environment for learning and gratifying memory formation. To satisfy our curiosity about his stored memories, when he was looking for a fitting letter or geo form to place it in the puzzle plane, we held in our hands but visible to him. When he saw the appropriate letter in our hand, he snatched from us and placed it on the puzzle. We have to assume that neuronal network have made specific designs that fit those designs outside his brain, those already fixed in the puzzle box. At that age, his visual input, motor movement and coordination, and a host of brain nuclei have been connected and synchronized to construct size, shape, color and letters designs to fit into one single concept, fitting. All the internal brain tools, meaning neurons, glia cells and molecules were there to get into action on signals from the nucleus. A single stimulus might have provided the necessary spark to create the whole complex brain response. We read books for learning, but it is the brain that constructs the patterns and designs to convert it into memory for storage and retrieval when necessary. Somehow throughout evolution, human brain cells have separated the human species, over other animals. Only human brain cells have been specialized to repair part of the body that feeds it. There are animals like the salamander that grows up a lost tail, but man does not regrow its bodily parts; he plans it and builds it in the laboratory. Next in his ambition to control nature, human brains make plans to leave our planet for another world. Only human brains can transcend its own existence and prepare its own body for a journey into the future.

TRILLIONS OF SYNAPSES

You and I have developed a brain with trillions of synapses capable of anywhere in the universe in time and space. While birds have developed its sight, a hundred times more powerful than humans, dogs and cats have developed their smell cells also more powerful than any human can; we have developed our brain like no other animal on this planet has. There are animals that can walk, fly and swim, but there is not a single animal in our planet that can build a car and drive it. An easier task would be making a pair of shoes for their feet. Some animals can be trained to perform specific task, but the capacity to store memories, put it into writing and using it when needed is uniquely human. Learning from our own memories either in writing, painting, music or just memory retrieval is uniquely human. It seems that learning from each other has been the great jump forward in our evolutionary process. Learning from each other requires trusting each other, and in trust we have the potential to care and empathize for humans. It is not by coincidence that our limbic system now, including the prefrontal cortex developed before the rest of the neo-cortex. The point is that we became a feeling organism that progressively developed a complex, rational and dynamic neo-cortex, which is a giant move forward. The distinctive quality of our brain is its capacity to create and modify its own designs for self-improvement and progress. An original idea can be constantly modified until the final product is satisfied. That is the job of the brain specialization and specification. All mammalians share a limbic system, but only human developed a complete neo-cortex finely wired to form, store, retrieve and use memory like no other animal has.

E.E.M. ON THE BEACH

Man has been traveling on land, sea and lately, on air for many years. We have looked and wonder at the vast sky with so many small lights-stars- incrusted in space for us to wish we could travel unrestricted and visit those faraway places. And, if we cannot do it by walking or flying, let our brains develop a module- a group of neurons, do the flying for us. My educational episodic memories, E.E.M. have excited neuronal circuits and pathways in my brain motivating me to further improve myself. It is a stimulation trigger that arouses compatible neuronal networks. Consequently, synapses get above normal excitation and promote similar circuits with consonant memories to join together and create a pattern or mental designs for a project that I want to work on. E.E.M. is long term memories, meaning it is stored somewhere in my brain, but I am not aware where they are stored. They are not bugging me or frustrating me when I am enjoying myself. A toxic or noxious memory would do that. E.E.M. will awaken up or reactivate association of memories that share something in common. This type of memory becomes the working horse in my brain looking for a common denominator to share a common goal. We could be out in the beach watching seagulls fly, glide in the air and float on the water waves. And, by association trigger, my experience while floating on air in a space capsule.

While I was watching seagulls float on water waves by the beach using my wrist watch; I had a paper, a ball point pen and notepad to keep track of seagulls A and B time floating on waves and flying. During this time, I was using my working memory to remember the minutes and seconds each seagull spent on water and on air, and writing it down.

For some unknown reason, I made a conscious association, at least it was my impression; that seagulls were gliding on the air as I was moving inside an energy capsule during my space traveling. Watching seagulls brought up to consciousness by association, a long past educational episodic memory. This seems to be a good example of working memory, short-term memory and long-term memory working together sharing a common goal. My visual working memory watching seagulls and other birds floating on the water waves, faded away temporarily, perhaps joining E.E.M. while both looked for already excited synapses that have stored similar visual stored memories. On previous pages in a different chapter, we said that E.E.M. has helped me isolate and weaken the noxious power of Delgo's memory. It might well be that E.E.M. neuronal network has become stronger to a certain extent, partially replaced toxic memories. Simply talking, new molecules were ordered by the neuron nucleus and formed dendrites bumps or protuberance, thus creating new synapses. There might be two processes involved during the formation of E.E.M. First, there might be over-excitation of synapses that after a significant time of neuron firing, become overly aroused requiring the formation of permanent memory. The permanent memory is dependent on the genesis of a new protein for storage and permanency of episodic memory. An extreme case of toxic episodic memory is post traumatic syndrome disorder or P.T.S.D. The fear stimuli kept the amygdale and related neuronal nuclei in the <u>on</u> position for a long time producing permanent anatomical changes in dendrites. This is an involuntary dendrite bump or protuberance.

"Studies employing PET and fMRI have shown that people with symptoms of PTSD have altered activity in the brain, primarily in the regions of the medial prefrontal cortex, thalamus, and anterior cingulated gyrus. This altered activity may facilitate and reinforce the brain's ability to recall specific traumatic memories, making it difficult to brake the pattern of negative memory recall."[54] Forty years of clinical practice back me up. You can go round those bumps and take away some of its toxicity, but the scar can flare up at any time under stress.

We have spent invaluable precious time of yours on memory stored in your brain. Mostly everybody has seen a picture of a brain

---

[54] The Brain, the Britannica Guide to, Running Book Publishers, Philadelphia, 2008, p. 267.

in books, television screen, magazine and periodicals. A close look at the whole brain, it seems to be broken into two parts or hemisphere, namely, right and left hemispheres. The two hemispheres are joined together from front to back by a stem or group of nerve tissue called corpus callosum that relays information on both directions. Some of the brain pictures show blood vessels covering the jelly like matter we call the brain. Unlike the mind, you can see and touch the brain. It is covered by a membrane that provides partial protection. The blood supplies the brain, among other things, oxygen and glucose, the main survival nutrients. During our lifetime, we get a lot of infections and are prescribed with medication to combat it. The virus and bacteria that get to us circulate in our blood system. Our ingenious brain developed a mechanism to protect itself from invading organisms. So, before the blood enters the brain, it is filtered out by the brain blood barrier. There are good reasons for the brain to develop a self-protective shield against microbes riding in our blood. If you get a wound or fracture in your body it will soon be repaired, but cells in the brain do not replicate and divide like with the rest of your body, except in the hippocampus and ventricles. The rest of this massive organ does not repair itself. If it would replicate and divided its own cells known to us as neurons, the memory stored in it will be gone. That means that you and us could not remember anything, and it would be a terrible problem for us.

ASTROCYTES AND
MEMORY FORMATION

Neuroscientists have discovered that brain cells named glia-cells are responsible for inspecting the blood from entering the brain. "Astrocytic end feet express glucose transporters…, water channels called aquaporins…, and a high density of potassium channels. The end foot specialization makes it highly probable that astrocytes play important roles in glucose uptake…, and in water and ion balance in the brain. The astrocyte mediated homeostatic functions are crucial for the stability of neuron function.[55] Once more, the above quotation comes to prove the crucial function of glia- cells in the brain complex function. Besides contributing to maintain the stability of the brain blood barrier, it has other equally important functions in the brain. We provided ample evidence that memory formation takes place at the level of synapses, the communication loci for neurons to talk to each other. Glia- cells that share the brain with neurons make up to 90% of all human brain cells. Among glia-cells, we have astrocytes that are involved in many things including to trigger the release of molecules required for establishing synaptic memory. Astrocytes make up over one hundred thousand of synapses in the hippocampus. Astrocytes tend to establish colonies like group thus, a monopoly of functions may be involved in establishing long-term memory. [56] We brought in glia-cells and specifically astrocytes because they seem to play a very important role in memory formation. They are engaged in triggering and releasing of calcium at presynaptic levels before the neurotransmitters is released into the synapse. As the

---

[55] Halton & Parpura, Glia-Neuronal Signaling, Kluwer Academic Publishers, Boston & London, 2004, page 3

[56] C.Hennenberger et al. Nature, Vol. 463, Issue #7278, 01/14/10, p. 169-70 + 232-235.

above quotation shows you, astrocytes are crucial for a normal hippocampus behavior. Without adding anything else, we would like to remind you of. H.M., whose hippocampus on both hemispheres was removed. He could not store nor retrieve any recent memory formed after his surgery.

GRAY MATTER

The gray matter shown in pictures of the brain corresponds to the brain densely packed neuronal cell bodies. In the cell body, you will find the nucleus with multiple sub-units performing a myriad of interesting and crucial functions. We can call this gray matter as the executive or decision making part of the neuron. However, we have to be careful to say this because the hillock that joins the axon to the cell body as well as the axon electro-genesis sodium /potassium pumps is indispensable for neuronal normal functioning. The axon is the long cable attached to the cell body (gray matter) by the hillock; it is covered by myelin. Myelin is the white fatty substance produced by a glia cell known as oligodendrocyte. A failure by the gene responsible for the synthesis of fatty matter can be fatal. Some of the most critical and painful brain diseases related to myelin deficiency are angio-trophic lateral sclerosis, (A.L.S), multiple sclerosis, cerebral palsy and Alexander's Disease.

WHITE MATTER

White matter fills nearly half of the brain. It consists of millions of cables, nerve fibers that connect individual neurons in different brain regions, like trunk lines of connecting telephones across the country. Recent discoveries in brain research that relate to our essay are the following: 1. when learning a complex skill, noticeable changes occur in white matter (glia- cells) functions; 2. on animals, shows myelin can change in response to mental experience; 3. a higher development of white matter structure, correlates directly with higher I.Q."[57] Higher I.Q means that we are smarter, that we are more intelligent than perhaps, our neighbor with a lower I.Q. It also means that we can pass city, state and federal test with higher score and get the job we are looking for. It may also mean that we can have more and better problem solving potential when the need arrives. A genetic failure to produce oligodendrocytes, a glia cell, can cause very serious health problems.

We have centered our essay on memory, learning, brain diseases, and disorders. Now we would like to introduce to you a case history of a person who has memorized more books with all its related bits of extra information that makes me wonder if we will ever be able to understand the brain. It is the story of Kim Peek. Born on November 11, 1951. His head and brain are highly oversized. Most strikingly, his brain has no corpus callosum, the large nerve tissue that connects to both brain hemispheres. Neurons in each hemisphere send their axons across the corpus callosum. It is how each hemisphere communicates with, share information and functions. For example, you are in the highway driving at 70 miles per hour and suddenly you see a car slowing down, the hemisphere that deals with space will inform the other hemisphere that it is time for you to move you right foot to the brake pedal. Of course, there are other steps taken before the

---

[57] Scientific American, March 2008, p. 54-61

executive part of the brain decides to reduce velocity or stop completely. Kim's brain lacks the connecting tissue between both hemispheres. Besides, Kim's cerebellum that is responsible for body balance, among other things, is smaller than usual with fluid covering the remaining space. It is what brain scans have shown on Kim's head. These abnormalities were brought with him from birth. Before you pass judgment on this special person, we beg you to read more about Kim's abilities. We have called him special for the memory storage he possesses. Kim began to memorize books, read to him when he was 18-month-old. It is not a printing error; it says 18-month-old. During this interview, most likely, early 2005, Kim had memorized 9, 000 books by heart and could read a page in 8 to 10 seconds. He knows all the area codes and zip codes in the U.S.A. He can identify hundreds of classical compositions with date, place and composers biographical details. A savant, he walks with a side gait, cannot button his clothes and has great difficulties with abstraction.[58]

We hardly know anything about his early life history, but whatever we learn from his brain scans and short personal interview by authors, D. Treffert and D. Christensen makes us wonder about the intrinsic works of the brain. In an intelligence test administered to him, he scored in the superior range in some areas while in other areas, he had to be placed in mentally retarded range of I.Q. From superior range of intelligence range to mentally retardation is a long way to measure. He has memories upon memories stored in his brain, but how they communicate with each other is, if possible, a challenge to him and to all of us. His brain is larger than most of us, but the connections to form patterns and multiple mental designs for problem solving and abstract thinking have been delayed, at least until now. He did not give up or surrendered to nature's accidents. He looked for the environment to compensate for his disabilities. At around age 50, he began to show newly developed skills. Socialization has been his greatest blessing. A Mozart scholar commenting on his thousands of facts stored in his brain's database reveals enormous intellectual

---

[58] Scientific American, 12/05, p. 108- 113.

capacity.[59] We have not been able to move out of memory and more memories because it is what makes us humans. Memories, learnings and feelings seem to be our basic human components that have built the separating bridge between us and the lesser developed hominids. In reference to Kim Peek, physical exercise, specially jogging and running for at least one hour a day, has amply proven to promote dendrite growth.

Researcher Li Huei Tsai at M.I.T. has shown us how a protein encoded by gene Cdk5rap2 (small 2) is linked to unusually small brain defects. This gene interacts with another protein, pericentrin, and malfunctioning of this protein is associated with abnormal brain size.[60]

---

[59] Ibid
[60] Nature, 5/13/10, Vol. 465, Issue# 7295, p. 145

SUBUNITS FROM THE OUTSIDE TO THE INSIDE OF THE BRAIN

Scientific researchers have repeatedly demonstrated to us that a rich environment, like healthy nutrition, home security, harmony among family members, no alcoholic beverages, no smoking and general good health during pregnancy is a precondition to have a healthy child. We excluded from the above statement, diseases secondary to genetic and epigenetic birth defects. Our environment has full of risks and health threats that we must avoid; we must be alert to air and water pollution, excess car toxic carbon dioxide emission, as well as multiple virus and bacteria. The last century left us with Aids or H.I.V., which we have not been able to break down and destroy. No less tragic are World War I and W.W. II when 26 million Russians died and five million Jews were sent to the crematoriums, gas chambers or just simply shot to death. All this carnage was provoked by a single wicked individual, Adolph Hitler and a group of fanatics who blindly followed him, the Nazi Party. Thousands of American soldiers died defending our country from Hitler and his gang. Besides the death of millions of innocent people, many more were left traumatized for life. Young boys and girls were separated from their parents and abused verbally, physically and sexually. Now at 2010, people with ages 70 and up gather together to support each other because those tragic-toxic–memories still haunt them. They confess how miserable their life; they have had to endure because there were individuals like Hitler, Stalin, Mussolini, Franco and the Japanese Military Elite. They were butchering people for the sake of gaining power and glory at the cost of human life. Many of us have vivid memories of relatives that either died or were wounded during W.W. II. These stored memories make us reflect on things we have done to ourselves.

Emmanuel fighting
Nazis in Europe

I remember my eldest brother who joined the U, S. Army and went to Europe to combat the German forces; he came back home in 1945. I was in the eighth-grade of the elementary school. He described to me the Nazi Party and its leader as a group of evil-minded people with no remorse or concern for human life. I dared not to tell him about how Delgo was torturing me because I considered my physical and mental anguish as insignificant compared to the suffering of the Nazi's victims. However, I cannot escape the pain when I see in the television screen, people with my age or younger sharing with humanity their tears, often openly crying, while revealing an open mental wound in their brain. It is the episodic memory we have been referring to in this essay that produced permanent structural brain changes in each one of us.

My brother spent a good piece of his lifetime sharing his experiences in war-torn Europe. He turned his attention to religion and philosophy attempting to teach brotherhood and love among men. He was advising people to watch out for violent, overly aggressive and power hungry individuals whose goal is lead us into wars. Many survivals of WW II, Korea, Vietnam, Iraq and Afghanistan have in their brain's painful episodic memories that prevent them from enjoying a happy life. We have shown you through extensive quotations from pioneer researchers of the brain, how new synapses become permanent and often dramatically change our behavior. The dendrite bumps or swelling carrying your toxic memory will be your life companion, torturing you if you decide not to do anything to reduce its power and decrease its grip on you.

EPIGENETIC MARKERS AND D.N.A.

We know that our genetic code, our D.N.A. holds the blue-print of what we are based on each of our parents inherited 23 chromosomes making a total of 46. We also know that crucial elements during our development, as well as some diseases, come about because our chemical components and its molecular and cellular structure have been modified. You may know how our body building blocks are composed of proteins, which are simple lines of peptides. Similarly, peptides are formed from amino acids in three letters (codon) of our genome code, namely: adenine, thymine, cytosine and guanine, briefly, A-T, C-G codify for an amino acid. For the sake of clarification, we take the three letters for the amino acid, glycine: GGU (actually four codons can codify for glycine) and additional molecules is added to this code. These additional molecules are called epigenetic markers. One of these markers is a methyl group. This specific group is composed of one carbon atom and three hydrogen atoms, which as a chemical formula is written CH_3. We must advance to you that epigenetic is <u>not</u> D.N.A. encoded. This additional methyl molecule in the D.N.A., meaning attached to the above codon for glycine will reduce gene expression. But, "It is linked to key developmental events."[61] Methyl is not the only epigenetic marker. During adolescent-hood, boys and girls undergo stressful physiological and psychological developmental events provoking physical and behavioral changes. We wonder if episodic memories like Delgo could have had provoked permanent structural changes in my brain. What if permanent changes occurred in critical location for learning and memory like the hippocampus and the prefrontal cortex heavily involved in deciding what we are and what we do. How about permanent episodic toxic memory in the prefrontal cortex decreasing its capacity to refrain the individual from antisocial and psychopathic behavior?

---

[61] Nature, 5/3/2010, Vol. 465, Issue 729r, p. 145.

Phineas Gage's accident would be a classical case to study based on his erratic behavior after the accident, but all our findings done with recent scanners are inconclusive. We do not know for sure which part of his cortex was seriously damaged. We have worked with patients with severe blows to their head but unfortunately, we did not have the necessary machines to follow up the case. However, we have seen television documentaries on returning war veterans from Vietnam, Iraq and Afghanistan whose behavior have led them to commit suicide and homicide. Combat experience is an extremely stressful and potentially dangerous situation for anyone to be in, regardless of the unit you belong to. We have not seen psychological testing and brain scan done on war veterans that come out with mental disorders that have had test done before and after discharge from service. Related to dendrites bumps for the formation of permanent structural changes, there is the possibility of using R.N.A. interference and other short R.N.A. molecules to prevent dendrites from forming or even dissolving the existing swelling within a limit time frame of the stressful experience. Researchers have mastered the science of the translation process at R.N.A. level to use its multiple different molecules as clinical tools to begin solving some of our mental disorders and diseases. Besides R.N.A., Craig Venter Institute has created a synthetic molecule with promising potential use in medicine that could be applied to brain disorders and diseases.

HOW CONSCIOUS WAS I OF DELGO LASTING PUNISHMENT ON ME?

Seemingly unrelated to our essay, but consciousness defined as electrical and chemical activities of a group of neurons in a specialized region of our brain, is related to anyone who has been subjected to repeated punishment. We can argue that for Delgo to have left such a permanent negative impression on my brain, I have to assume that I must have been overly conscious of myself. I had to be overly conscious because the cluster of neurons and pathway responsible for keeping an animal, an organism awake and alert, had to be turned on, all the time. Lizards are known, among other things, for their long life on planet earth and primitive brain. They are awake and alert in order to jump in the next meal. However, to attribute human consciousness to an alligator is out of the question. Every day of the week, I entered Delgo's classroom expecting to have some kind of ridicule and punishment. I was very conscious of myself and what lies ahead of me. Both groups of neurons, those in-charge of keeping me alert and those responsible for consciousness were turned on at all times. They had to be <u>on,</u> to avoid Canis Delgo from pointing me out as sleeping during his class time. I am aware that I may sound and look obnoxious to you; but by forcing my brain to turn the switch on at all times, Delgo was imposing on my neuron's abnormal excitability and potential brain damage. Yes, I can anticipate your thoughts: the amygdale, hippocampus and prefrontal cortex were the groups of cells most exposed to overexcitation and damage at the time. Among the alternatives I had at the time, were to walk out of school and go home to escape him, but I did not. I cannot say that I enjoyed him torturing me. In retrospect, my brain must have made numerous designs to block incoming stimuli from him. The electrical-chemical activities that took place in many of my brain neurons were responding to the fear and disgust I felt for him. Perhaps, we can argue that I was getting habituated to his cruel behavior, and I had gotten numb to his cruelty. We have treated patients under this condition. There must be a threshold, a limit of tolerance of action before the brain's self-protective system goes into effect. Slaves and individuals that have been physically and emotionally abused or exploited and survived, must have had built a protective emotional shield within themselves to prevent further deterioration. I figured this out because the brain's self-protective devices would not continue to build permanent dendrites swellings or bumps culminating in brain malfunctioning and subsequent death.

On the other hand, if I became a passive recipient of abuse and my brain stop responding to pain, I believe my brain cells would stop or decrease their normal firing capacity. When a situation like this one actually takes place, learning and subsequent positive memories are seriously compromised. However, in my case, I was determined not to give up. It seems that there was a match or contest between Delgo's wicked thoughts and actions and my brain's surviving tools. There is no doubt that I was growing up a depressed and emotionally insecure youngster during most critical years of my life. I believe it hit me most during my 15th to 16th years of age, when the opposite gender takes a prominent place in shaping our masculinity. We wanted to show our masculinity, and being ridiculed by another person in front of someone you are interested in, is extremely painful. Growing up as a depressed youngster which was being ridiculed and verbally abused in front of the girl you are in love with, is not a small matter. If I could not defend myself, how could I defend her? My self-esteem must have been very, very low. However, as I pointed out in the previous pages, the girl I was pretending had a brother who hated Delgo for similar behavior. It was a kind of incentive for me to hold on and keep on track. I did not do well academically and my worse grades were from Delgo. I figured out that the brain areas mostly compromised during those years in school with Delgo are: the amygdale, which is heavily connected with learned fear. The amygdale has been described as the main door to the limbic system and emotions. Also heavily compromised has to be the hippocampus, which you have already been introduced to its role in memory and learning. Not obnoxious, but teachers like to repeat part of the last lesson before starting a new one; remember H.M.'s case for memory retrieval and storage? The prefrontal cortex including the orbital and medial frontal cortices added to the hypothalamus that is the regulatory nucleus for stress and arousal. The triumvirate consisting of the hypothalamus, pituitary and adrenal gland must have been ready to secret adrenaline and cortisone at the slightest stimuli from Delgo. It has been shown over and over again that an excess of it, as an end product of chronic stress, kills brain cells in the hippocampus. If cells died in the hippocampus, your ability to learn or even to retrieve what you have learned is seriously compromised. When I look back into my years of torture with Delgo's hand, I repeat aloud to myself, "You have done very well, Jose."

IN THIS ESSAY WE HAVE USED THE WORDS CONSCIOUSNESS AND MEMORY AD-INFINITUM

Memory and mind are often used interchangeably not only by lay people, but by supposedly well-educated individuals as well. In antiquity, the mind was a domain of the soul. Most religions in planet earth, antiquity is alive and well-cultivated by its followers. During my personal experience with Delgo, I hope I have not left a doubt about memory formation and its location within our brain. I do not believe you will not agree with me that Delgo`s the main contributor to my long-lasting toxic memories. It is was his name and punishment that I remember the most among my teachers. It was his name and humiliating experience that was spontaneously, unwanted and vividly retrieved in the midst of my depressive episodes. When I was supposed to be enjoying a pleasant conversation, it was his memory that interrupted the normal and healthy flow of chemicals and thoughts in my brain. If for the sake of argument, you claim that it was alright for him to call my attention that I was not properly dressed for his classroom, you might score a point. But, he picked on me even when we moved near the school and there were no more cattle for me to care for before going to school. Secondly, my girl friend and fellow students came over to me verbalizing in disbelief how Delgo was getting away with unfair discipline and punishment. You can try to explain my toxic episodic memory in many different ways, but if you take Delgo out of the picture, there would be nothing else for you to work on. The remaining memories would fall in place beautifully like in Brandon's puzzle. Of course, there are problems in transcription, translation, snps and epigenetic, but it belongs into another group of memories, if at all.

On the location of memory formation, I selectively quoted for you, Nobel Laureate's researchers as well as the less privileged, but nonetheless brilliant scientists that are synaptic and memory enthusiasts and researchers. You have the expertise of psychologist and teachers that can be witness to many empirical datum and information leading to brain plasticity and memories. In vivo and in vitro experiments with lower animals, researchers have demonstrated, not the memory itself, but the bumps formed after a memory is completed. Memory formation as we have attempted to conceptualize is a complex process well above our imagination. "Memory studies, for instances, usually focus on the effect of disrupting a particular gene or use a psychological test to try to determine how memories are made. But neither of those types of experiment reveals which neuronal circuits are activated during the memory-making process."[62] If you allow me to go back to Delgo, I would say that when he provoked abnormal neuronal synaptic excitability in my brain, thus, promoting the arousal of postsynaptic receptors to force the nucleus of the neuron to activate transcription and translation involving R.N.A in protein expression, my toxic memory had the beginning of its genesis.

---

[62] Science News, January 30, 2010, p. 19.

DELGO'S TWO FACE PERSONALITY

When you have your eyes, ears, skin and guts set in; you do not mind sending your brain unpleasant perceived signals from the outside. Meaning from Delgo, a large part of your brain was seriously compromised. My functional brain was trying to bring order and normalcy to form an appropriate response to a threatening external stimulus. How would you respond if you were confined in a cage-like prison cell with a psychopath like Delgo? You may consider it an exaggeration of the facts, but in above pages I told you I confronted parents whose children had been verbally and physically abused by the same person. Their assessment was hardly different from mine. The main problem for us to really challenge Delgo was his two face personality. For some member of the community, he was the law and order man they needed. Good discipline was maintained at school; no smoking or drinking of any kind was allowed in or around the school, and above all; most parents dared not go to school with any sort of complaint. If a parent could not keep Delgo's standard of behavior, school attendance was not enforced after the sixth-grade. On the surface, it was a happy and just school system. However, similar systems were established in Germany, Spain, Italy, Japan and Russia under totalitarian governments.

By the time I left school in May 1946, many W.W.II veterans had come home after defeating Hitler's Army in Europe and Japanese in the Far East. Delgo`s image and authority in the community slowly began to decline. He dared not challenge recently discharged war veterans who were still experiencing combat mental backlash. On several occasions, we saw war veterans still wearing their fatigue uniform coming to school visiting the School Principal and classroom teachers without Delgo's interference. His almost daily stroll to nearby stores was stopped suddenly. Discharged soldiers showing their regimental and division insignia were patrolling the area during the day while attending school in a nearby town at night. What a welcoming sight it was for me. My nightmare was over although, the toxic memory remained inside me for many years more.

MEMORY IS POWER

What is a memory if not a power to retrieve and bring to consciousness the past experience of your life? There are many types of memories besides the one we have dedicated in this essay. There are conscious and preconscious memories that come and go without much effort from our part. They may respond to a joke from a friend or a painting from Picasso, Goya, Monet or Van Gogh. For some of us, a stroll around the boardwalk watching seagulls float on the waves is mentally soothing and artistically inspirational. We find a comfortable place to rest and our brain will do the rest, busy doing patterns and designs coming deep from inside us. Our brain forms those designs from sensorial stimuli converted or translated into memories for us to enjoy. Walking along the boardwalk, a girl passes by and, if from nowhere, the image of Maggie replaces that of the seagull. All is a memory. During the summer, we go out to the country where flowers are in abundance, and its perfume take my brain back to Maggie. All is a memory. Passing list to my life history book, and… Maggie writes the last quotation.

MAGGIE, A STROLL IN PARADISE

Unlike Delgo, Maggie's memory is a stroll in paradise. They are incomplete memories, but memories nevertheless. My marriage, children, trips overseas, friends, parties and accomplishments are, "Memories." What am I if not for memories? Whether conscious, preconscious or unconscious, it all belongs to me. Sweet, sour or neutral; my memories were made in my brain by my brain, and it is for me to make the best of it. Although still working on it, incomplete memories are in the process of satisfactory completion. People, learning and memories have given me a house full of treasures. Whether it is an unpleasant visual, auditory and taste memory, I have learned to either change or replace it with pleasant ones. When the road is excessively threatening and rough to walk, I have the freedom to choose my next step. I do not have the power to change anyone, but I can choose who and what I like to listen to. I have learned a lot since I left Delgo. This type of learning is stored inside my brain. My memories can also be stored in CD, books, magazines and in the brain of another person. I have learned from many people; what I have learned from others and made my own through assimilation, I hope I can teach someone else. Memories made all of us human beings able to care for and love each other. Memories are written in stones and caves, a thousand years ago. Egyptians left their memories written in hieroglyphics; Babylonians in cuneiform and the Mayan memories are still in the process to fully understand its meaning. We feel blessed for having a brain with power to retrieve pleasant memories of days long gone and be able to enjoy it as if it is happening here and now. Memories have made us rational multicultural organisms capable of critical thinking and a highly developed cultural and scientific society.

REWARD CENTERS IN MY BRAIN

Maggie replacing seagulls, a stroll at the boardwalk or simply medi-tating in my backyard tends to bring synchronicity during my brain neuronal firing. Meditation and researchers have shown us spikes of brain activity in the left pre-frontal cortex resulting in increased focus and attention, improving performance on cognitive tasks... It also helps reduce anxiety disorders, high blood pressure, insomnia and depression."[63] We have named the neurotransmitter dopamine as the pleasure producing chemical in the endogenous reward system. Also, involved in this naturally occurring pain and anxiety decreasing sys-tem is the endorphin producing groups of brain cells. Endorphin comes from the words endogenous and morphine. Anatomically, we know at least have three brain areas that are actively engaged in producing a kind of euphoric states; it is a joyful and happy subjec-tive feeling where reality may dictate in us something quite different. Reality in this multiple universe is no easy to explain. Not in the sophisticated and extremely well connected and developed region of our brain, but in an archaic and primitive region of our developing brain. Between the ponds and midbrain, you will find a small nucleus named ventral tegmental area. As you might have guessed already, it uses dopamine as a neurotransmitter. Further complicating our wellbeing, this archaic area has a significant and powerful projection to another reward group of brain cells committed to make us feel extra happy. Its projection is in nearby limbic system called "nucleus accumbens." All animals without the well-developed cortex enjoy this primitive for joy brain network. You do not have to be human to let your body enjoy itself under the influence of dopamine, endor-phins and other proteins with pain relieving properties that are found in the brain. However, useful and engaging to stimulate humans to multiply its own species, among other benefits, the ventral tegmental

---

[63] Scientific American, Mind, Feb- March, 2009 p. 62-63.

area extended itself far beyond its surrounding into the cortex. But, like a tree extending its roots to the most fertile ground, these small ventral tegmental nuclei originally in the midbrain, covered the limbic system through the nucleus accumbens, and now, projected into the most sophisticated of all our brain regions, the prefrontal cortex, particularly, in the left hemisphere. Now, the smartest of all animals have three endogenous reward systems: the ventral tegmental area, nucleus accumbens and prefrontal cortex. Illegal drugs brought into our country like heroin, cocaine, marijuana and more, hijack our brain reward against pain and cut short our happy life on planet earth.

THE REWARD SYSTEM
AND ADDICTION

The smartest of all animals learned how to use and abuse those three centers for joy and pleasure for self-recreation. Either by accident or experiment, man came across herbs like marijuana, cocaine and heroin that placed him in another state of consciousness for his own benefit, at least, it is what he claims to be doing. The dopaminergic neurons along with endorphins served man well for many years. However, his experiment with himself and the above-mentioned herbs have brought self-destroying and society problems. All addictions tend to use, at least, the nucleus accumbens at all times. Controlled experiments done in animals including man, has proven this group of cells activated in all addicts. My years of practice have shown how otherwise responsible, honest and hardworking parents have used their children for pedaling drugs on the street. In a worst-case scenario, their children were rented as an object of sex to bring in money for drugs. There was no remorse or sense of guilt for what they were doing. The good maternal and paternal memories of care and love for their children were ignored or simply put aside. They did not care about themselves or the children as long as drugs and a subjective feeling of joy was coming to them. What they did not understand, was that the drugs they were consuming had hijacked not only their three-pleasure system, but their entire brain and life. Memories from family, parents and friends became dormant and unresponsive. Recreation, education and parental love were thrown overboard and addiction had taken over. It is an outside war against the self. Memories like Delgo hijack the amygdale, sympathetic system, thalamus, hypothalamus- pituitary and adrenaline axis, but you can fight back. When an addiction sets in, your prefrontal cortex is no longer at your service, it is part of the conspiracy against your surviving instinct. Neuroscientist and professor, Jeanette Norden writes,

"In animals you have certain drugs that can hijack the system (reward system); it happens in humans as well. Drugs like cocaine artificially stimulate these same pathways…We know that individuals who are drug addicts …talk about how they get those whole-body euphoria which is so intense…this intense feeling of elation that occurs from taking the drug…(they)will refuse to eat or drink and die if not forced to eat. [64] The naturally appearing reward system when altered by outside stimuli can become very destructive. You must have read about serial killers and the joy they experienced when they murdered anyone. Likewise, you have a man that sexually abuse small children, not because of sexual satisfaction in itself, but the joy of destroying a defenseless and innocent human being. Often, these individuals were subjected to mental and physical abuse and torture. Their brain wiring was altered or modified in a way that their behavior is a threat to themselves and society. Instead of pleasant memories to enjoy life, their toxic episodic memories are in a homicide trip looking for a prey to feed their thirst for pain on innocent victims.

---

[64] Jeanette Norden, Understanding the Brain, The Teaching Co. Virginia, 2007, p. 167-73

MEMORIES AND DECISION MAKING

The making of a decision is based on stored memories which are experiences and available to an individual. The decision is subject to time, place, need, urgency and usefulness, among others. A decision may involve many circuits and pathways between neurons and groups of neurons at different location of the brain. Most likely, a functional neuronal network for normal communication is involved. If for some reason, communication between two or more brain nucleus is not possible or compromised; the decision is cancelled, modified or postponed. For example, Ivan, a person with moderate Parkinson's disease goes with his family to a zoological park, and his family decides to take a look at the wild cats 200 yards away. Ivan forgot to take his medication with him and his tremors, especially his legs and hands signaled him that he should be heading home for his medication. A dopamine deficit was behind his decision. However, had Ivan had no Parkinson's disease, but had been jumped on by a wild cat 30 years ago, when he was on a hunting trip celebrating his 18th birthday, most likely, it would be an entirely different decision. The wild small tiger had stuck his canine teeth on Ivan's left arm. He was taken to the hospital and the surgeon did the necessary repair. Ivan became extremely apprehensive of cats, hunting and wild animals. In this scenario, Ivan's family invitation to take a look at wild cats has a different color, meaning a different behavioral and physiological response. Did the family know that Ivan had a traumatic memory of a wild cat? Would the sight of a tiger similar to the one that had jumped him 30 years ago have an impact on his decision now? Would the tiger sight trigger the amygdale into action and Ivan feel his heart beats go faster and louder? If Ivan's heart was running like a wild horse, his blood pressure gone up and his sweat glands betraying his apparent coolness; would a memory thirty years ago become suspicious for provoking an abnormal physiological response in a healthy and rational individual that took a stroll around a zoological park? Ivan and his family were aware that the tiger was behind bars in a safe cage with no possibility of harming him. However, despite his family and Ivan's awareness that there was no possibility of the tiger jumping the cage and attacking him; he was not spared of an abnormal spike in his blood pressure. Would you characterize Ivan's experience as a synaptic over excitability or a permanent dendrite anatomical change? Have you had experience-memories that can be classified as causing a permanent change in your brain? Would you like to develop an assessment scale with numerical values 1-5 based on intensity of pain? Five would be the most painful memory. Can you pair the most painful memory with one of your pleasure memory? How long did it take for the first time you tried it? How about the second time? Could you change the end result and become the victor instead of the victim? Try it several times and write it down; keep it in a safe place and read it to yourself as often as you can.

ON AND OFF GENES

Actually, turning on and off memories is done by the brain itself when appropriate and necessary. Perhaps, post-traumatic stress disorder is the most commonly known case when the brain turns itself off. It is not our intention to become obnoxious, but the amygdale and its related pathways during the traumatic episode, remains turned on much longer than the victim wishes. If the victim freezes or collapses during the traumatic episode, most likely she or he become unconscious and have a blackout. Blackouts do not simply occur in our brain only. I remember two blackouts in New York City and one that covered in darkness the whole North-Eastern United States. During those three episodes, it was explained to us that the system shut itself up. It was an overconsumption of available energy. Well, when your brain and my brain are overloaded, when an input overflows our alert system, our brain turns itself off. Internally, our brain has on and off genes that respond to protein responsible for synchronization of brain activity. Outside brain stimuli like caffeine and nicotine make your brain more excitable releasing neurotransmitter like glutamate into synapses. Glutamate, unlike GABA, another neurotransmitter, is an excitatory chemical in our brain. As we have said earlier, our brain is a complex and very sophisticated organ that has learned to master its surrounding. It has more cells synapses than stars in the sky. Therefore, you can expect an organ like the brain to develop safety system for its own protection. During a cortical period, brained-derived neurotrophic factors turns on the nucleus basalis- the part of the brain that allows us to focus attention-and keeps it on, throughout the entire critical period.[65] We do not think we will be aware of all neural activity taking place just above our neck, right in our head. We must explore and learn about our brain actions. There are

---

[65] Norman Doidge, The Brain that Changes itself, Penguin Books, New York, 2007, p. 79-80.

millions of people in America and around the world that cannot fall asleep because their brains do not shut off. Some cannot fall asleep because of psychological problems such as P.T.S.D. The alert system we referred to above, do not turn off and stays on high alert despite the harm that it brings to itself. The brain does not seem to be aware that the threat to itself is over. During the trauma, the network connecting regions and organs in the brain must have suffered a sort of short circuit. Remember Delgo, it was not easy to put down the power of that painful and toxic memory. It goes without mentioning it that the brain itself may bring genetic failures that prevent it from normal functioning. Making a decision with a brain full of past painful memories compounded by gene failures is a challenging task for anyone. We need generous, trusting, loving and caring persons around us.

ADDICTION AND MEMORIES

When a person experiences the stimulus of an addictive substance like heroin or cocaine the brain's reward system gets the message that it is a pleasure producing product. A group of dopamine releasing nuclei in the ventral tegmental area located near our most archaic brain region system, the brain stem, sends its projection to another "feel good" group of neurons, the nucleus accumbens. This group of brain cells love to feel high. Whenever you are enjoying anything, meaning your body has turned on pleasure, the nucleus accumbens is on. Laboratory work done on animals including humans, reveal that the nucleus accumbens is flooded with dopamine and the subject under study is in a high pleasure scale number. Individuals that run away from the hardship of daily life, as well as traumatic memories bugging his/ her brain, turn to drugs for temporary relief. The problem with this seemingly easy solution is that the person becomes addicted to the pleasure producing substance. This person will be pulled into further consumption of drugs by her/his own brain in need and calling, "give me more drugs." It is not an episodic traumatic memory amenable to psychotherapy; it is a brain hijacked by a drug. The victim or drug addict will need more cocaine or heroin to compensate for the relatively "small" pleasure received from dopamine produced in the ventral tegmental area. Further complicating and aggravating the high experience of the addict, genes are turned on implicating the transcription process of brain cells. For a temporary feeling of pleasure, the person compromises basic processes of cell metabolism, division and multiplication. It is very sad for me to read newspapers saying, we consume more illegal drugs than any other society in the world. My experience of almost forty years working with addicted individuals brings me very deep and intense pain.

The addicted individual's capacity to think clearly and make rational and logic decisions is seriously compromised, if not obliterated altogether. He or she must have their daily doses of drug.

The addict feels his life will soon end if he does not have his drug. Drug addicted parents are willing to sell their children for drugs; they are willing to rent their children for sex to buy drugs. The drug addicted person main and only concern is his next shot of drugs. There are millions of individuals around the world hooked on drugs not because they want to. In most cases, it is poverty, ignorance, lack of will power and the power of drugs producers and distributors. Lack of job opportunities in their neighborhood attract youngster into urban areas just to fall prey to drug peddlers. Near forty years of experience in the largest city of the United States qualifies us, to make valid statements. Attempts to learn more about the brain reward system, scientists working on mice behavior have found that shining a blue light on the nucleus accumbens activated connections to the amygdale. The goal was to have a stimulus activate the long known pleasure center, accumbens. Whenever there is a chance for an exaggerated feeling of joy, the nucleus accumbens welcomes it. Enjoying a good book must turn on the ventral tegmental area send its projections into the accumbens. A happy reading will activate neuronal pathways and networks consonant with the book experience. You will store new memories making you wiser and more intelligent. The person may become more intelligent because synapses become sensitized with the potential of developing into long-term memories. You are adding new experiences that enhance your own life expectation. We are emotions and memories. Even very small children react negatively to tarantulas, they have never seen before. Those child memories are stored in the collective unconscious of thousands of years ago. Tarantulas, snakes, darkness and a child left alone produces memories of a long past that until now, provokes goose pimples on us.

The D.N.A.-R.N.A.- Protein Dogma in Memory Process

Shape determines protein function. Proteins are the building blocks of our body. From the hair on our head, nails on our toes to newly formed cells in our hippocampus-center for learning and memories-you will find protein doing its corresponding job. Proteins are linear chains of polymers known as amino acids. There are many amino acids, but only 20 are needed to form our proteins. Our D.N.A., our genetic molecule that contains our life code, only code for 20 amino acids to build your body and mine. Actually, in the nucleus of our great chromosome molecule, genomic D.N.A. is wound around histone proteins forming nucleosome, which are packed into beads; the package is known as chromatin. Nucleosomes are the basic single units in the chromatin structure. This D.N.A. histone protein package must be tightened together to prevent unwinding as well as extra-cellular interference with normal gene function. Among the gene's functions is the forming of protein and memories, especially long-term memories need protein to be formed by our brain cells. This delicate packaging of D.N.A. and histone proteins purpose culminating in chromatin is to make sure that genes are properly and perfectly packaged. An error in the packaging would result in a gene failure or change leading to a dysfunctional protein or no protein formation. Scientists have discovered brain diseases and pointed out the culprit as poor D.N.A. protein packaging. In the surface, it seems that the gene is the villain. However, the problem is more complicated that appears to be. Epigenetic plays a leading role in brain diseases.

Cells are programmed to become special cells from early in the embryonic stage of development during the blastocyst phase, which is around 4-5 days after conception. Despite embryonic cells predisposition for specialization at an early stage of development, cells carry the instructions to express every structure in our body. Scientists doing research to find a solution to religious and moralists' group objection to use embryonic stem cells as tools for medical purpose led to the discovery of I.P.S. The technique of Induced Programmed Stem cells takes adult stem cells and revert its programming to embryonic stem cells. For example, a liver or hematopoietic cell can be reprogrammed to have all the potential of an embryonic stem cell. The adult stem cell is taken from the patient to be used for the treatment of his own disease. This process is supposed to remedy the objections of the above groups that object to the use of stem cells as a treatment modality and research tool. Our great D.N.A. molecule with its companion of histone proteins and guest of honor, the mitochondria, has developed useful tools for proper functioning and survival. Cells turn genes <u>on and off</u> when needs to be done. Cells can use their own machinery to form and attach a molecule to a gene it needs to shut off. One of those chemical markers, the cell uses to silence a gene is a methyl group (one carbon atom and three hydrogen atoms). Another cell chemical marker is an acetyl group. The cell can attach it to a protein, gene or D.N.A. itself, altering its functions and activity in many ways.

A Quandary

We said that the cell developed many of these tools throughout evolution. Now, we are facing a quandary. For the cell to develop chemical markers, does it require memory during the developing stage of synthesis (forming) the marker molecule? Should we call it a genetic code not involving memories, as we have used it during our essay? It sounds very convenient to us, but it seems we are side-stepping a crucial issue. Earlier, we said that even highly specialized cells carry instructions to form all types of structures in our body. Is it memory or just a genetic code developed throughout millions of years of evolution? Let's say, a monkey left the tree it had called home for many years and walks away from it to an open and inviting plain with big yellowish bananas. He eats them until his belly is round, almost like that of a hippopotamus. He heads home but does not take any bananas with him. The following day, he returns with three or four of his friends. The bounty of bananas lasted for about four months. We believe you will agree with us that the banana memory took hold of him; he remembered there was a tasty food in the other side of the plain. How about the changes that his feet began to go through? His feet, like his arms, were used to climb and hang from tree branches, not for a long walk on the plain. The monkey began to take short strolls along the plains while an imaginary banana was bugging his head. In time, the monkey spent more time searching for the bugging banana while neglecting his sweet home, the tree. His feet changed or you can say, adapted to walking. The banana memory changed his brain; walking the plain changed his feet. The stimuli he received on his feet while walking the plain were registered in his brain. How would you characterize those experience registered in both his feet sensory cells and brain cells? Is it fair to say that the feet change is simply evolutionary adaptation while the banana memory is cognitive growth? Someone bugging us would say that we are completely wrong. He or she would argue that whether feet or hands changes, it is the brain that ultimately is responsible for any change in the body. Another brain fan would comment that brain plasticity provides for changes in bodily adaptation. She would even argue that arms, legs, ears, nose and everything else in our body, exist for the use of our brain. In other words, cells, and in particular, brain cells carry the instructions to shape and reshape itself when needed. Remember, our brain, our organism learns and grows through memories. What makes us different and superior to all other forms of animal life is our ability or potential to store memories from learning. The same obnoxious reminder, "If it were not for memories, you would not be reading this essay." Well, if scientist Craig Venter was able to make a synthetic bacterium in a couple of years; then, it is not surprising for a single cell 3 ½ billion years old to make all necessary tools for self-survival. Doctor Craig Venter and his team of researchers must have accumulated many memories in their brain to successfully achieve the dream of the many, but the glory of the few.

MEMORIES STORED OUTSIDE MAN'S BRAIN

Ever since man learned to transfer his thought-memories into symbols, his cognitive potential took a gigantic step forward. Anthropologists have found caves in France and Spain decorated with animal drawings many thousand years old. More advanced civilizations in Mesopotamia, Egypt, Mayan in Mexico, Incas in Peru and Ecuador have used stones, cuneiform, hieroglyphics, and other forms of communication to leave us their memories, their collective wisdom. From the cave messages, man went into pyramid-building, obelisk, statues, mausoleums, and even stone cities (Maccu Pichu) on top of rocky mountain to secure and protect their heritage, accomplishment and memories for future generations. From traditional storytelling, man created a more reliable, accurate and lasting form of conveying his thoughts, his memories. Man learned to apply heat and possibly pressure to create clay tablet to inscribe their memories for us. Besides, he left us stone carvings, sheep-skin, papyrus and metals recording his glorious memories of past events. He recorded names of Kings, Queens, wars, religious manifestations and beliefs as well as nature phenomena like hurricanes, earthquakes, volcanic eruptions, floods and extra-terrestrial visiting. Man interpreted all these events according to his mental development and understanding. These memories became part of the collective unconscious of a tribe, nation, country or empire. The Old Testament, the Egyptian Book of the Dead, the Popol Vuh or Mayan Book of the Dead as well as Hindu and Chinese ancestor books have contributed to our understanding human memories imprinted in many forms.

HOMO-SAPIENS

Whether our human species emerged or sprang from a single paternal hominid or multiple hominids in different places and time on planet earth, our homo-sapiens has striven to pass on to us the best of his accomplishments and memories. Our ancestors have built on rocks, challenging the forces of nature like the ferocious and destructive force of the wind, water flooding and earthquake to deliver their memories as complete and intact as possible. Our ancestors compete with each other, to protect us for the best of their creation. They build for eternity, so that the coming generations could learn from it and improve it if the necessity arose. The message was to avoid distortion and commit to memory their wisdom so we can move forward and prosper. Greek, Sumerian, Persian, Egyptians, Hebrews and Mayan historians have received the call of our ancestors of long ago and passed it on to us today. Herodotus, the Greek historian left us in writing the Trojan Epic. The Hebrews left us their religion in a narrative fashion and the Romans left us besides their language, the Coliseum, aqueducts, palaces and formidable roads. Marco Polo left his trip to China and Christopher Colombo the narrative epic of the discovery of America. And in America, Neil Armstrong took our memories to the moon. We have no doubt that our memories will travel to other solar systems and galaxies. We are sending electromagnetic signals into the outer space to alert other forms of intelligence and, possibly, establish communication with them. We have sent proves into space with similar objective. Our memories once limited to cave drawings, are evolving and searching its origin. Life on planet earth is only 3 1/2 billion years old; our universe after the Big Bang is about 17 billion years old. Astrophysicist Steven Hawkins has verbalized, he has no doubt of the existence of other forms of intelligence in our universe. In fact, it would be a terrible waste of celestial space and time if we were the only piece of land in the universe to have life and intelligent animals. Our quest for intelligence in outer space points to our desire to join with our creator, whatever it might be. In the end, it might a journey in a circle.

THE MOLECULAR BASIS OF MEMORY, INCLUDING DELGO

Leaving dualism, the universe aside and emphasizing humanity, mid-twenty century psychologist and scientists have embarked themselves in a challenging task: find the molecular basis of memory in the brain. Among the American pioneers in the search for the brain mechanism of learning and memory, we had Karl S. Lashley at Johns Hopkins University. Lashley took a psychology class with John Watson, considered by many enthusiasts as the father of Behaviorism in America. Initially, Pavlov classical conditioning became the basis of his research. Lashley, on the other hand, became interested in the physiological response of the reaction during each experiment. His objective was to trace the conditioned reflex pathway through the central nervous system. Related to our traumatic experience with Delgo, Lashley wanted to find where and how did my punishment go into my brain. As you might have guessed, it involved the amygdale, hippocampus, thalamus, autonomic system (sympathetic) hypothalamus, pituitary, adrenaline gland, limbic system and undoubtedly, the pre-frontal cortex. With the whole brain engaged in surviving that physical and emotional assault, localization of my memory was out of the question. Besides, at the time, behaviorists were only interested in observable behavior during the experiments.

Lashley was aware that the observable behavior was just the tip of the iceberg if he wanted to explore on a deep level of learning and memory. Observing how the wind of a hurricane crushes down houses does not tell you the whole story of how and what forces make up a hurricane. In 1956, we witnessed the force of 145 miles per hour hurricane wind in a wooden house and it was not fun when part of the house disappears with the wind. Lashley, instead, went deeply into the brain to find the locus for learning and memory. He was one of the American scientists to challenge the brain to reveal its mysterious and challenging functions. His ultimate goal was to find the location in the cerebral cortex where the engram or memory trace was stored. The tracing of memory in the brain had taken steam, "In 1949, Canadian psychologist Donald Hebb proposed that learning and memory are based on strengthening of synapses that occurs when pre-and postsynaptic neurons are simultaneously active."[66](J.M. Schwartz, the Mind and the Brain, 2003) We believe in the saying, "Neurons that fire together, wire together." The molecular foundation of memory attracted worldwide attention of neurologists, psychologist and scientists of many orientations. At Columbia University in New York there was Eric Kandel already mentioned in this essay. In many ways, his work has set the pace for further research in this area.

---

[66] James M. Schwartz, The Mind and the Brain, Regan Books, N.Y. 2002, p. 107.

PLASTICITY

Researchers were learning to open the black box, the brain that had been the subject of interesting speculation and divisive arguments for millenniums. We learned that the brain was not a rigidly wired box, but a dynamic and plastic wonderful organ that we needed to explore. Archaic and ridiculous religious, and popular cult beliefs considering the brain as god given organ for man to keep away from, it (,) was challenged by courageous human behavior researchers. Experiments with animals and accidents in humans have proven without any doubt that neurons can take over functions from tissue and organs not originally theirs. It seemed that neurons were rearranging themselves according to received external stimuli. The brain as a whole became an object of scientific study. The time for speculation on its nature and relationship to a supreme and untouchable creator has given way to sophisticated laboratory staffed by brain specialist. For example, when researchers created small retina lesions in an animal's eye, the cortex did not receive input from that area, but input from the retina surrounding the lesions were processed by the brain. This type of plasticity takes place in young and adult brain. The brain, besides dynamic and adaptive, can form a bodily map that fits external stimuli as well as reshape already made maps. Not only group of cells take over extra-territorial function, but plasticity refers also to hemispheric activity as well. A lesion in one brain hemisphere may have a corresponding group of cells on the opposite side of the brain that if activated, can respond and take over additional functions.

Doctor Ramachandran
Insight

A prominent neurologist from the University of California, V.S. Ramachandran with an extensive clinical experience and research on brain lesions patients became excited and highly motivated about discoveries done on neurons taking over functions not originally wired to do so. The end specification of neurons was not rigidly wired as generally thought of and taught at medical school. There was room for neurons to move around and grab territory that was not occupied by its legitimate owners. It was also true with neurogenesis. Neurogenesis was almost a taboo subject among professionals in the field of medicine just two decades ago. Doctor Ramachandran addressed a scientific meeting in 1993 and alerted the audience that his human amputees experience cerebral cortex reorganization similar to findings in monkeys. His research and clinical practice with phantom sensations moved to the brain neuroplasticity, a step further into rehabilitation medicine. Jeffrey M. Schwartz states, "Neurons that originally fired in response to stimulation of a now-missing body part look for new work, as it were, and instead respond to peripheral neurons that are still in the game."[67] There are millions of Americans and worldwide individuals that have suffered stroke, arm and leg amputation as well as other physical disabilities that can benefit from this newly discovered plasticity of the brain. After all, it seems that Donald Hebb firing neuron was able to turn on the curiosity of many brain scientists around the world for our benefit. There is no doubt in my mind that plasticity in my brain helped me overcome Delgo's toxic memory and write this essay. How my brain was able to overcome Delgo's memory anatomical change is too late to find out, but definitely, Delgo is out of combat now. Besides the tools I

---

[67] Jeffrey Schwartz, op cit, p. 186

designed and practiced while destroying him, I rediscovered myself more powerful than ever before. My hidden strength was identified, strengthened and consolidated in one healthy and happy person. I set realistic goals to achieve during my lifetime. I learned to question my behavior objectively and make appropriate adjustments when necessary. I learned to choose authors of books that would help me most while keeping my own self-identity. I like to remember one of my professors saying, "Make sure you chew it well before you swallow it, otherwise you may get indigestion."

MEMORY RETRIEVAL

Memory seems to be a simple process of storage and retrieving a sensory perception. For most people means remembering a past personal event, a folklore narrative of a national hero, outstanding family members achievement, childhood episodic memories, imagined identification with an ideal self like Che Guevara, Robin Hood, Daniel Boone and many others. We can bring up (retrieve) many of our memories with great ease, but some need some effort on our part to bring it forth. We also have "forgotten" memories and repressed memories. Repressed memories are held back in our brain by self-protective forces or mechanisms that have guaranteed our survival for thousands of years. Besides the above mentioned memories, we have visual and acoustic memories that are very particular and personal for each individual person. In addition, we have musical memories that might be a universal language, but a Mozart Sonata may not mean much to a Jibaro Indian in the Amazon jungle. Memories are colored by each person's religious and cultural orientation as well as his/her emotional and mental development. For example, a believer receiving 20 slashes on his back while practicing-experiencing- a religious ritual procession might be a horrifying sight for some while it could be the cleansing of sins for another. Pain and pleasure is like a circle, it all depends where, when and how you jump into it. A person's voice, face and behavior may help us bring forth memories that were not easy to remember. The brain makes multiple associations from different stimuli often creating scenes, designs and even behavior that seems alien to the person. Perhaps an extreme case in this area is a bed dream or nightmare. The dreamer experiences objects, animals, birds, fish and people he never seen before, not even in science fiction. The most common dysfunctional designs of the brain are hallucinations and delusions. The brain creates a false reality for the victim or patient. The patient behavior is dictated by the false reality created by his brain. When a schizophrenic has a visual hallucination, he/she will be responding to that reality, which is a false construction by the brain. However, for the schizophrenic, the false reality created by his brain is a reality guide that he believes in and follows as it dictates. His conversation with a person you cannot see is a reality for him. People are not totally wrong when they say, "It is all in his mind, except that I would correct them and add, in his brain."

MEMORY LOCI

Neuroscientists have discovered that memories are formed and stored (Eric Kandel), but the brain decision to store specific memories to a location or locations in the central nervous system is in need more research. Among the many memories that get into our brain, we have forgotten memories, among others, that influence our behavior in many ways in the present time and place. These memories may be actually forgotten because they were of temporal nature. We can say that more pertinent and useful memories activate relevant synapse overriding less important issues. We may compare it to our daily struggle for survival. There seems to be a competition among memories. Some memories may be taken over, perhaps, by an emotionally loaded memory. In a different scenario, memories are rewarded more favorably than others. Our opiate system)endorphins) can push aside, so to speak, less relevant memories. On the other hand, in our case for instance, I do not remember what happened in the classroom on a daily basis, but Delgo's memory was my unwanted and aversive companion. The brain fear system had taken over the feeling part of me and the pleasure system was subsided by Delgo. Further elucidating our attempt to find a location for our specific memory, we like to mention stroke victims, especially in the left hemisphere involving Broca and Wernicke areas. Broca victims are unable to verbalize their feelings while Wernicke seem to modify or alter the experience or memory. Unlike Broca and Wernicke, and similar accidental cases that trigger depression, the etiology of my depression must have a mother child relationship. Delgo's memory anatomical change in my brain, Delgo can be loo at as a survival threat and challenge to me. My oldest sister and mama tiny had done an excellent job on me. The challenge that Delgo had imposed on me mobilized my survival strength. Despite much abuse and humiliation, u had an audience at home to show off my innate conquering abilities and ambition. It was Delgo the teacher that I had to conquer and defeat. The more

I enjoy pedaling him, the weaker his image. At times it seems to be a laughing joke. I celebrated my 90 birthday a few months ago. My memories have changed quantitatively, but specially, qualitatively for the better. My brain has served me well. When a synapse or two is damaged by someone like Delgo, we have trillions of synapses capable of taking over.

Amen.

www.ingramcontent.com/pod-product-compliance
Lightning Source LLC
Chambersburg PA
CBHW030509210326
41597CB00013B/846